I0033924

# PESTICIDE CHEMISTRY AND TOXICOLOGY

Dileep K. Singh

# eBooks End User License Agreement

Please read this license agreement carefully before using this eBook. Your use of this eBook/chapter constitutes your agreement to the terms and conditions set forth in this License Agreement. Bentham Science Publishers agrees to grant the user of this eBook/chapter, a non-exclusive, nontransferable license to download and use this eBook/chapter under the following terms and conditions:

1. This eBook/chapter may be downloaded and used by one user on one computer. The user may make one back-up copy of this publication to avoid losing it. The user may not give copies of this publication to others, or make it available for others to copy or download. For a multi-user license contact permission@benthamscience.org

2. All rights reserved: All content in this publication is copyrighted and Bentham Science Publishers own the copyright. You may not copy, reproduce, modify, remove, delete, augment, add to, publish, transmit, sell, resell, create derivative works from, or in any way exploit any of this publication's content, in any form by any means, in whole or in part, without the prior written permission from Bentham Science Publishers.

3. The user may print one or more copies/pages of this eBook/chapter for their personal use. The user may not print pages from this eBook/chapter or the entire printed eBook/chapter for general distribution, for promotion, for creating new works, or for resale. Specific permission must be obtained from the publisher for such requirements. Requests must be sent to the permissions department at E-mail: permission@benthamscience.org

4. The unauthorized use or distribution of copyrighted or other proprietary content is illegal and could subject the purchaser to substantial money damages. The purchaser will be liable for any damage resulting from misuse of this publication or any violation of this License Agreement, including any infringement of copyrights or proprietary rights.

**Warranty Disclaimer:** The publisher does not guarantee that the information in this publication is error-free, or warrants that it will meet the users' requirements or that the operation of the publication will be uninterrupted or error-free. This publication is provided "as is" without warranty of any kind, either express or implied or statutory, including, without limitation, implied warranties of merchantability and fitness for a particular purpose. The entire risk as to the results and performance of this publication is assumed by the user. In no event will the publisher be liable for any damages, including, without limitation, incidental and consequential damages and damages for lost data or profits arising out of the use or inability to use the publication. The entire liability of the publisher shall be limited to the amount actually paid by the user for the eBook or eBook license agreement.

**Limitation of Liability:** Under no circumstances shall Bentham Science Publishers, its staff, editors and authors, be liable for any special or consequential damages that result from the use of, or the inability to use, the materials in this site.

**eBook Product Disclaimer:** No responsibility is assumed by Bentham Science Publishers, its staff or members of the editorial board for any injury and/or damage to persons or property as a matter of products liability, negligence or otherwise, or from any use or operation of any methods, products instruction, advertisements or ideas contained in the publication purchased or read by the user(s). Any dispute will be governed exclusively by the laws of the U.A.E. and will be settled exclusively by the competent Court at the city of Dubai, U.A.E.

You (the user) acknowledge that you have read this Agreement, and agree to be bound by its terms and conditions.

**Permission for Use of Material and Reproduction**

**Photocopying Information for Users Outside the USA:** Bentham Science Publishers grants authorization for individuals to photocopy copyright material for private research use, on the sole basis that requests for such use are referred directly to the requestor's local Reproduction Rights Organization (RRO). The copyright fee is US $25.00 per copy per article exclusive of any charge or fee levied. In order to contact your local RRO, please contact the International Federation of Reproduction Rights Organisations (IFRRO), Rue du Prince Royal 87, B-I050 Brussels, Belgium; Tel: +32 2 551 08 99; Fax: +32 2 551 08 95; E-mail: secretariat@ifrro.org; url: www.ifrro.org This authorization does not extend to any other kind of copying by any means, in any form, and for any purpose other than private research use.

**Photocopying Information for Users in the USA:** Authorization to photocopy items for internal or personal use, or the internal or personal use of specific clients, is granted by Bentham Science Publishers for libraries and other users registered with the Copyright Clearance Center (CCC) Transactional Reporting Services, provided that the appropriate fee of US $25.00 per copy per chapter is paid directly to Copyright Clearance Center, 222 Rosewood Drive, Danvers MA 01923, USA. Refer also to www.copyright.com

*Dedicated To My Students*

# CONTENTS

# About the Author

Dr. Dileep K. Singh is an Associate Professor in the Department of Zoology, University of Delhi, Delhi and is working on pesticide toxicology, bioremediation of pesticide molecules, soil microbial ecology, and soil health and soil fertility. He has about seventeen years of teaching experiences in Postgraduate Departments of the University *i.e.* in the Department of Zoology and Department of Agrochemicals and Pest Management and has similar years of research experiences. He has completed nine research projects funded by IAEA/FAO, NATP-CGP (World Bank), ICAR, DBT and DST. Four ongoing research projects are funded by DST, DBT, IAEA/FAO and DU-DST Purse Scheme.

He has published 27 international papers in journals like J. Agricultural Food Chemistry, Environmental Science and Technology, Chemosphere, Canadian J. Microbiology, Biodegradation, J. Environ. Science and health and Bull. Environ. Contam. Toxicol. Impact factor of these journals ranges between 0.9 to 4.36. About 9 national papers are published in journals like Pesticide Research Journal, Indian J. of Microbiology and Indian J. of Entomology. He has been an Executive Editor of Indian J. of Microbiology since 2006. He has been elected Treasurer, Association of Microbiologists of India (2011-2013). Reviewed many international research papers *i.e.* FEMS Microbiology Letter, J. Environmental Science and Health (B), Bull. NRC, Applied Soil Ecology, International Journal of Analytical Chemistry *etc.* and book on microbiology. About 84 abstracts were published in proceedings of conferences, symposium and workshops. He has published two popular books on pesticides in Hindi and has contributed two chapters in book. He has been associated in writing chapters on Pesticide in e-book by NISCAIR (CSIR), New Delhi, India (2005). E-Chapters on Toxicology of insecticides and "Insecticidal method of pest control" are available on the website http://nsdl.niscair.res.in/, posted on 28th September, 2007. He is also a member of working group for creating e-content for ACPM constituted by NISCAIR, New Delhi , India (2006) Member of Content Advisory Committee for Biology, constituted by MHRD New Delhi for the preparation of e-Biology Book (2006) on http://sakshat.gov.in/

Ten students have obtained their Doctorate Degree under his supervision and six students are working for the same. Four Post Doctorate students have also worked with him. About 22 students have obtained their M. Phil. Degree under his supervision and about 25 students of M.Sc. (ACPM) have submitted their projects. He is a member of The National Academy of Sciences, Allahabad. India (2008), Pesticide Society of India, the Association of Microbiologists of India, and Entomological Society of India. His research group is working on pesticide residue analysis, soil microbial ecology, soil health and microbial bioremediation of pesticides. He was an elected member of Academic Council (2002-2006) of University of Delhi for two terms and had four years administrative experiences as Resident Tutor in one of the PG hostels of the University. He is a member of many committees in the university and in other organizations. More information about author is available on website http://people.du.ac.in/~dksingh/

# FOREWORD

I am delighted to introduce my student and the author of this e-Book, Dileep K. Singh, an Associate Professor, in the Department of Zoology, University of Delhi, India. He has been teaching 'Pesticide Toxicology' at master level since 1996. He has contributed many research papers on environmental pesticide toxicology in peer reviewed journals with good impact factor. I know him personally and appreciate him for his sound knowledge in pesticide toxicology. He conceived the idea to write a book on pesticide toxicology somewhere in 2006, since then he has been working on it. I have seen his book on 'Pesticide chemistry and toxicology', which is informative and deals with different aspects of toxicology and chemistry associated with the pesticides. I specially appreciate his efforts to include a chapter on 'Pesticide Metabolism', which provides the information about the pesticide degradation in very lucid manner and well explained using figures with target site for pesticide metabolism.

Pesticide Metabolism is typically a two stage process. These are Phase I reactions and Phase II reactions. Phase I reactions, normally add a functional (polar reactive) group to the foreign molecule which enables the phase II reaction to take place. These reactions are catalyzed by the cytochrome P- 450 group of enzymes and other enzymes which are associated with endoplasmic reticulum. Phase I reaction includes microsomal oxidation and extramicrosomal metabolism of pesticides. Phase II reactions are conjugation reactions and involve the covalent linkage of the toxin or phase I product to a polar compound. In general, conjugated products are ionic, polar, less lipid soluble, less toxic and easily excretable from body.

I hope that this e-Book will be very useful to the students of master and graduation level and also to the people working in different fields of pesticide toxicology.

*Professor H.C.Agarwal*
F.N.A.Sc., F.N.A.A.S. *(Retired)*
Insect Biochenistry & Toxicology
Department of Zoology
University of Delhi
India

# PREFACE

As a teacher, I have faced several questions put forward by my students on toxicology such as dose response relationship, carcinogenicity, teratogenicity, mutagenicity, metabolism of toxic chemicals and their behavior in the environment. These queries initiated me to write a book on "Pesticide chemistry and toxicology", which will provide some answers to their questions. In this e-Book, there are eight chapters with subtopics; each is dealing with answers of very specific questions on toxicology. I have tried to pin point the discrepancies raised in my mind as toxicologist and tried to pen down here. I hope this e-Book will be helpful to the students of pesticide toxicology and scientists/teachers working in this field. I am thankful to my colleagues for their continuous support in writing this e-Book and NISCAIR (CSIR, India) for according the approval to use my two chapters hosted on their website. Thanks to Bentham Science publisher who have made publishing this e-Book possible.

*Dileep K. Singh*
Delhi, India

# ACKNOWLEDGEMENT

I am thankful to my students for providing me valuable information and queries related to this eBook. I am thankful to my wife Ms. Saroj Singh, my daughter Ms. Shagun Singh and my son Mr. Kartikay Singh for their constant support in writing this eBook. I am highly indebted to Bentham Science publishers for providing me an opportunity to disseminate this knowledge to my students in the form of an eBook. I am highly thankful to different sources that are used as information in this eBook to fill the gaps in knowledge, these are mentioned at different places; without acknowledging them my liabilities towards the society and science will not be completed.

*Dileep K. Singh*

# KEYWORDS

Pesticide ; Insecticide ; Classification of pesticides, Toxicology ; $LD_{50}$; $LC_{50}$; Dose response relationship; Pesticide Metabolism ; Phase I reactions; Phase II reactions; Insecticide formulations; Insecticide resistance; Antidotes.

# CHAPTER 1

# Pesticide Chemistry

**Abstract:** Pesticides are biocides especially designed to kill, repel, attract and mitigate the organisms, which are nuisance to humans and their agricultural and hygiene activities. The pesticides may be the organic molecules or inorganic synthetic molecules or the biopesticides. The toxicological activities on pests depend on their chemical structure and different life stages. These pesticides can be classified as organochlorine compounds, organophosphorus compounds, carbamates, pyrethroids and neonicotenoids. In organochlorine compounds, the number and position of Cl in molecule decides the toxicity. They are nonpolar and lipophilic in nature. Organophosphorous pesticides (OP) are neutral ester or amide derivatives of phosphorous acids carrying a phosphoryl (P-O) or thiophosphoryl (P-S) group. OP pesticides are identified by single characteristic *i.e.*, they act by inhibiting cholinesterase enzyme. Carbamates are anticholinesterase inhibitor pesticides and are synthetic derivative of physostigmine, also known as eserine, which is a principle alkaloid of plant, *Physostigma venenosum,* Calabar bean. Chemically, they are esters of carbamic acid, $HOOC.NH_2$, with insecticidal (and related) properties. Pyrethroids are esters formed by the combination of two acids i.e Chrysenthemic acid and Pyrethric acid, with three alcohols namely Pyrethrolone, Cinerolone and Jasmolone. They are nerve poisons and affect the nerve axon, causing repetitive discharge of nerves which results in eventual paralysis. The *neonicotinoids,* are the newest major class of insecticides, derived synthetically from nicotinoids. Mode of action of neonicotinoids is that they act as agonists at the insect nicotinic acetylcholine receptor (nAChR). In this chapter, we will study the chemistry of different group of pesticides.

**Key Words:** Pesticides; History of pesticides; Pesticide classification, Organochlorine pesticides, Organophosphorus pesticides; Carbamate pesticides; Pyrethroids; Plant origin pesticides; Biopesticides; Fumigants.

## 1.1 INTRODUCTION

A **pesticide** (pest + i (many words are formed with connecting 'i' ) + cide , L. *–cida,* it is from Latin word *caedere* means 'to kill', therefore the meaning of pesticide is to kill the pest) is a substance or mixture of substances used for preventing, repelling, mitigating or destroying any pest. A pesticide may be a agrochemical substance, organism (such as a bacteria and fungi), which is used against pest/pests. Pests include insects, mites, nematodes, molluscs, birds, mammals, plant pathogens, weeds and microbes, that compete with humans for food, agriculture commodities, shelter, act as a vector and cause a nuisance.

Pesticide is a broad term, it includes acaricides, insecticides, fungicides, herbicides, rodenticides, nematicide, molluscicides, ovicides, piscicides and bactericides. Many chemicals are also included as pesticide though they are not necessarily killing the pest nor their name ends with suffix "cide", the examples are chemosterilants, attractants, repellents, desiccants, defoliants and plant growth regulators. Many biopesticides are also included under this broad heading.

Pesticides are usually chemical substances and its active portion is known as active ingredient (AI), which is usually formulated as BB (block bait), CG (encapsulated granule), DP (dust able powder), dust, EC (emulsifiable concentrate), GR (granules), GB (granular bait), OL (oil miscible liquid), SP (soluble powder), SL (soluble concentrate), WP (wettable powder) and WS (water dispersible powder). Many of these are diluted with water before their use to target sites.

Method and pattern of pesticide usage also differ from each other depending on the target organism, their life stages, physiological status and prevailing environmental conditions.

## 1.2. DIFFERENT PESTICIDES

Pesticides are classified on the basis of their chemical structures, mode of action, method of entry into the body and target organisms. As navigating the major groups of pesticides, these may be named as Acaricide, Avicide, Algaecide, Antifeedants, Bactericide, Bird repellent, Chemosterilants, Fungicide, Herbicide, Insecticide, Insect

**Dileep K. Singh**
**All rights reserved - © 2012 Bentham Science Publishers**

Attractants and Repellents, Molluscicides, Nematicides, Rodenticides, Virucides and miscellaneous (Table **1**). Here, we will focus our information on Insecticides, Fungicides, Herbicides and Bio-pesticides.

| History of pesticides : Its development and usage. | |
| --- | --- |
| **PERIODS** | **PESTICIDE DEVELOPMENT AND ITS USAGE** |
| **BC** | |
| **8000** | Beginning of the Agriculture. |
| **2500** | The Sumerians used sulphur as an acaricide and insecticide. |
| **1200** | China used the chalk and wood ash to control the insects. They also used the plant extracts for the control of stored grain pests and arsenic sulphide to control human lice. |
| **1000** | Homers refer the use of sulphur compounds. |
| **320 -325** | Chinese use ants in citrus groves to control caterpillars. |
| **100** | The Romans apply hellebore for control of rats, mice and insects. |
| **AD** | |
| **70** | The use of sulphur as an insecticide. Arsenic, soda, olive oil was used for the treatment of legumes. |
| **900** | Chinese use arsenic to control garden insects. |
| **1600 -1700** | Introduction of botanicals such as pyrethrum, rotenone, derris and tobacco leaf infusion. |
| **1765** | The first organic insecticide was nicotine, it was applied in its natural form as crushed tobacco leaves for the control of aphids. |
| **1800** | Petroleum, Kerosene, Creosote and Turpentine were introduced as insecticides. |
| **1860** | Paris green, which is copper salt of arsenic was used to control colorado potato beetle, codling moth and other leaf-eating insects. |
| **1873** | DDT was first made in laboratory by Otto Ziedler. |
| **1880** | Lime Sulphur used in California against San Jose Scale insect. |
| **1892** | Lead arsenate was introduced as one of the most effective inorganic insecticides for the control of pests such as gypsy moth, apple maggot, and various soil insects. |
| **1900** | Sulphur, arsenicals, fluorides, soaps, kerosene and various botanicals such as nicotine, rotenone, pyrethrum, sabadilla and quassia were used as insecticides. |
| **1939** | The discovery of insecticidal properties of dichlorodiphenyltrichloroethane (DDT) by Paul Muller of J.R. Geigy Company, Switzerland. It was first used to control malaria and typhus by the Western allies during World War II. |
| **1940** | BHC insecticidal properties were discovered in France and England. |
| **1940** | Synthesis of organophosphorous insecticides was subsequently done on a world wide scale (starting from Germany). Three of these compounds HETP, Parathion and Schradan attributed to Gerhard Schrader, were extensively used. |
| **1944** | Phenoxy acetic acids were discovered as first selective herbicide e.g. 2,4-D |
| **1949** | Captan, first dicarboximide fungicide was introduced. |
| **1950** | American insecticide carbaryl was introduced. Malathion was introduced as safest organophosphorus insecticide. |
| **1951** | First carbamate insecticides are introduced. |
| **1949 to 1970** | The development of a number of synthetic pyrethroids. |
| **1972** | *Bacillus thuringiensis* (Bt) registered as insecticide. |
| **1976** | The first synthetic pyrethroid was introduced and since then they have become the second largest class of insecticide used today. |
| **Recent developments in** | These include the insect growth regulators (IGR), such as chitin synthesis inhibitors, juvenile hormone mimics, ecdysone agonists, pymetrozine, and other novel agents such as pheromones, |

| pesticide usage | *Bacillus thuringiensis*, avermectins, formamidines. |
| | Insecticides with novel mode of action such as Imidacloprid, Buprofezin, Benzoylphenyl urea, Cyromazine, Pyrrole insecticides. |
| | 122 nations, including US signed a treaty to phase out completely POPs including DDT. |
| December, 2000 | Many pesticides are withdrawn or banned for use due to toxicity and persistence in environment. |

Pesticides are classified in one of the two ways, either in terms of chemical characteristics or according to their mode of actions. Sometimes, we also classify them according to their action on target organisms.

**Table 1.** Different pesticides and their target organism (Latin suffix "-cide, -cides" means to kill or to cause death)

| Pesticides | Derivations | Target Organisms |
|---|---|---|
| Acaricide/ miticide | Gr. *Akari* means mites or ticks | Mites or ticks |
| Algaecide/ algicide | L. *Alga* means sea weed | Algae |
| Avicide | L. *Avis* means bird | Birds |
| Bactericide/bacteriacide | L, Bacterium | Bacteria |
| Biopesticide | Bio+pesticide | biological pest control |
| Fungicide | L. *Fungus* means mushroom | Fungi |
| Herbicide | L. Herba | Weeds |
| Herpecide | Gr. Herpeto | Reptiles |
| Insecticide | L. *Insectum* means cutup or divided into segments | Insects |
| Larvicide | L. Larva means young juvenile | Larva of insect |
| Molluscicide | L. *Molluscus* from *Mollis* means soft | Snail and slugs |
| Mycocide | L. *Fungus* means mushroom | Fungi |
| Nematicide | L.*Nemat* means roundworm | Nematodes |
| Ovicide | L. Ovum | Eggs |
| Piscicide | L. *Pisces* means fish | Fish |
| Rodenticide | L. *Rodere* means to gnaw | Rodents |
| Virucide | L.Virus | Virus |

## 1.3. PESTICIDES CLASSIFICATION

### 1.3.1 Chemical structure

#### 1.3.1.1. Main Groups

1. **Organochlorine compounds:** DDT, BHC/HCH, Aldrin, Endosulfan, Heptachlor, Methoxychlor, Chlordane, Dicofol.

2. **Organophosphorus compounds:** Parathion, Malathion, Monocrotophos, Chlorpyrifos, Quinalphos, Phorate, Diazinon, Fenitrothion, Acephate, Dimethoate, Fenthion, Isofenfos, Phosphamidon, Temephos,Triazophos.

3. **Carbamates:** Aldicarb, Oxamyl, Carbaryl, Carbofuran, Carbosulfan, Methomyl, Methiocarb, Propoxur, Pirimicarb.

4. **Pyrethroids**: Allethrins, Deltametrin, Resmethrin, Cypermethrin, Permethrin, Fenvalerate, Pyrethrum.

5. **Neonicotinoids:** Acetamiprid, Imidacloprid, Nitenpyram, Thiamethoxam.

### 1.3.1.2. Other Groups

1. **Organotin compounds:** Triphenyltin acetate, Trivenyltin chloride, Tricyclohexyltin hydroxide, Azocyclotin.

2. **Organomercurial compound:** Ethyl mercuric chloride, Phenyl mercuric bromide.

3. **Dithiocarbamate fungicides:** Zineb, Maneb, Mancozeb, Ziram.

4. **Benzimidizole compounds:** Benomyl, Carbendazim, Thiophanate methyl.

5. **Chlorphenoxy compounds:** 2,4-D, TCDD, DCPA, 2,4,5-T, 2,4-DB, MCPA, MCPP.

6. **Dipyridiliums:** Paraquat, Diquat.

7. **Miscellaneous :** DNOC, Bromoxyl, Simazine, Triazamate.

### 1.3.1.3. Other Classifications of Pesticides

1. **Aliphatic compounds:** Methyl bromide, Malathion, Glyphosate, Aldicarb, EPTC, Maneb.

2. **Aromatic compounds:** 2,4-D, Diuron, Carbaryl, Permethrin.

3. **Heterocyclic ring compounds:** Nicotine, Captan, Benomyl, Atrazine.

### 1.3.1.4. Pesticide Classification based on mode of Action

1. **Nerve poisons**: Organochlorine and Organophosphorus pesticides and Carbamates.

2. **Anticoagulants**: Warfarin.

3. **Juvenile hormones**: Azadirachtin, Fenoxycarb, Methoprrne, Hydroprene.

4. **Antifeedents**: Neem, Citrus derived limonoids and their synthetic derivatives.

5. **Repellents**: Permethrin, Neem oil, Citronella oil.

### 1.3.5. Pesticides Classification Based on Pesticidal Action

1. **Stomach insecticide**: DDT, BHC/HCH, Methoxychlor, Lead arsenate, Paris green, NaF.

2. **Contact insecticide :** Chlordane, Aldrin, Nicotine, Parathion.

Many other methods are also available to classify the pesticides, which is beyond the scope of this book.

## 1.4. CHEMICAL CHARACTERISTICS OF PESTICIDES

### 1.4.1. Organochlorine Pesticides (OC)

Organochlorine pesticides (also known as chlorinated hydrocarbons in which one or many hydrogen atoms have been replaced by the chlorine) are primarily insecticides with relatively low mammalian toxicity, fat soluble and normally persistent in the environment. Many chlorinated hydrocarbons have the ability to accumulate inside the body due to their lipophilic nature. Their main characteristics are,

1. Presence of carbon, chlorine, hydrogen and sometimes oxygen atoms including a number of C- Cl bonds. Number and position of Cl in molecule decides the toxicity.

**2.** Presence of cyclic carbon chains including benzene ring.

**3.** Lack of any particular active intra-molecular sites.

**4.** They are nonpolar and lipophilic (soluble in fat) in nature and have a tendency to concentrate in the lipid rich tissues, thereby causing its bio-concentration, and biomagnifications at different trophic level in the food chain.

**5.** Chemically uncreative, therefore highly persistent in the environment, resistant to microbial degradation.

Organochlorine groups were first used as pesticide in 1940s. Between 1945 to 1965, organochlorines were used extensively in agriculture and forestry, in protecting the wooden buildings and humans from a wide variety of insect pests. After awareness that these compounds are highly persistent, legal action has been taken to phase out this class of insecticides. It includes DDT, Lindane, Endosulfan, Aldrin, Dieldrin, Chlordane, Heptachlor and Endrin (Fig. **1**).

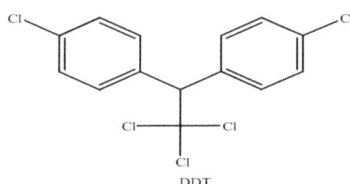

IUPAC : 1,1,1-trichloro-2,2-bis(4-chlorophenyl)ethane

FORMULA: $C_{14}H_9Cl_5$

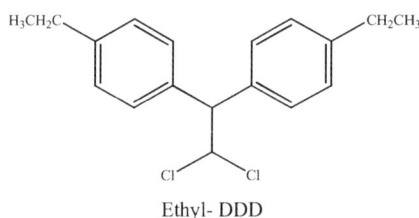

Ethyl- DDD

IUPAC: 1,1-dichloro-2,2-bis(4-ethylphenyl)ethane

FORMULA : $C_{18}H_{20}Cl_2$

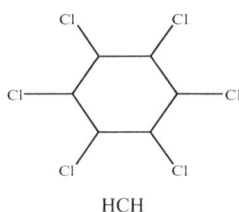

HCH

IUPAC: 1,2,3,4,5,6-hexachlorocyclohexane

FORMULA: $C_6H_6Cl_6$

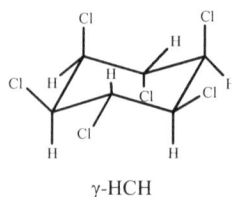

γ-HCH

IUPAC: 1α,2α,3β,4α,5α,6β-hexachlorocyclohexane

FORMULA: $C_6H_6Cl_6$

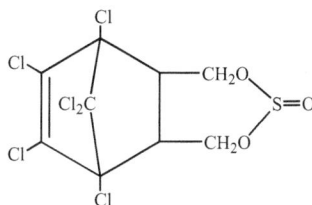

Endosulfan

IUPAC: 1,4,5,6,7,7-hexachloro-8,9,10-trinorborn-5-en-2,3-ylenebismethylene sulphite or 6,7,8,9,10,10-hexachloro-1,5,5a,6,9,9a-hexahydro-6,9-methano-2,4,3-benzodioxathiepine 3-oxide

FORMULA: $C_9H_6Cl_6O_3S$

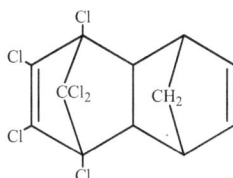

Aldrin

IUPAC: (1R,4S,4AS,5S,8R,8aR)-1,2,3,4,10,10-hexachloro-1,4,4a,5,8,8a-hexahydro-1,4:5,8-dimethanonaphthalene

FORMULA: $C_{12}H_8CL_6$

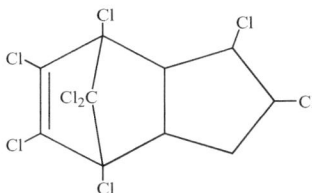

Chlordane

IUPAC: 1,2,4,5,6,7,8,8-octachloro-2,3,3a,4,7,7a-hexahydro-4,7-methanoindene

FORMULA: $C_{10}H_6CL_8$

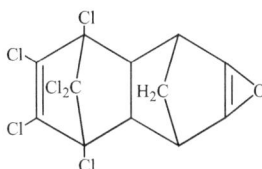

Dieldrin

IUPAC: (1R,4S,4AS,5R,6R,7S,8S,8aR)-1,2,3,4,10,10-hexachloro-1,4,4a,5,6,7,8,8a-octahydro-6,7-epoxy-1,4:5,8-dimethanonaphthalene OR 1,2,3,4,10,10-hexachloro-6,7-EPOXY-1,4,4a,5,6,7,8,8a-octahydro-ENDO-1,4-exo-5,8-dimethanonaphthalene

FORMULA: $C_{12}H_8CL_6O$

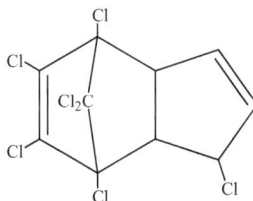

Heptachlor

IUPAC: 1,4,5,6,7,8,8-heptachloro-3a,4,7,7a-tetrahydro-4,7-methanoindene

FORMULA: $C_{10}H_5CL_7$

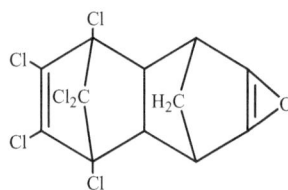

Endrin

IUPAC: (1R,4S,4AS,5S,6S,7R,8R,8aR)-1,2,3,4,10,10-hexachloro-1,4,4a,5,6,7,8,8a-octahydro-6,7-epoxy-1,4:5,8-dimethanonaphthalene OR 1,2,3,4,10,10-hexachloro-6,7-epoxy-1,4,4A,5,6,7,8,8a-octahydro-EXO-1,4-exo-5,8-dimethanonaphthalene

FORMULA: $C_{12}H_8CL_6O$

**Figure 1.** Structure of different organochlorine pesticides.

### *1.4.1.1. Chemical classification of organochlorine pesticides*

|   | **Types** | **Examples** |
|---|-----------|--------------|
| 1. | Cyclodiene compounds | Aldrin, Dieldrin, Endrin, Heptachlor, Isodrin, Endosulfan,Chlordane. |
| 2. | Halogenated aromatic | DDT, Kelthane (Dicofol), Methoxychlor |
| 3. | Compounds | Chlorobenzylate. |
| 4. | Cycloparaffins | HCH, Lindane. |
| 5. | Chlorinated terpens | Polychlorcamphenes, Polychlorpinenes. |

## 1.4.2. Synthesis of Organochlorine Compounds

### *1.4.2.1. Cyclodiene compounds*

The Diels-Alder reaction is used to synthesize the cyclodiene compounds, which is governed by the Woodward-Hoffmann Orbital Symmetry rules. Reaction of hexachlorocyclodiene with cyclopentadiene will give Chlordene, which on further chlorination of the adducts, it gives the mixture of two isomers, α and β chlordane (Figs. **2 & 3**).

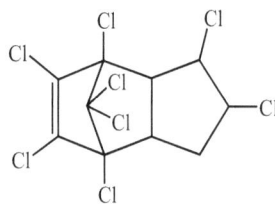

Chlordane

CHLORDANE : 4,7-methano-1H-indene, 1,2,4,5,6,7,8,8-octachloro-2,3,3A,4,7,7a-hexahydro-

**MW**: 409.778640 G/MOL, **MF**: $C_{10}H_6CL_8$

**Figure 2.** Structure of Chlordane.

### *1.4.2.2. Halogenated Aromatic Compounds*

Benzene and substituted benzene do not react appreciably with chlorine and bromine in cold, but in presence of "halogen carriers" such as iodine, iron, pyridine and aluminium amalgam, the reaction is fast.

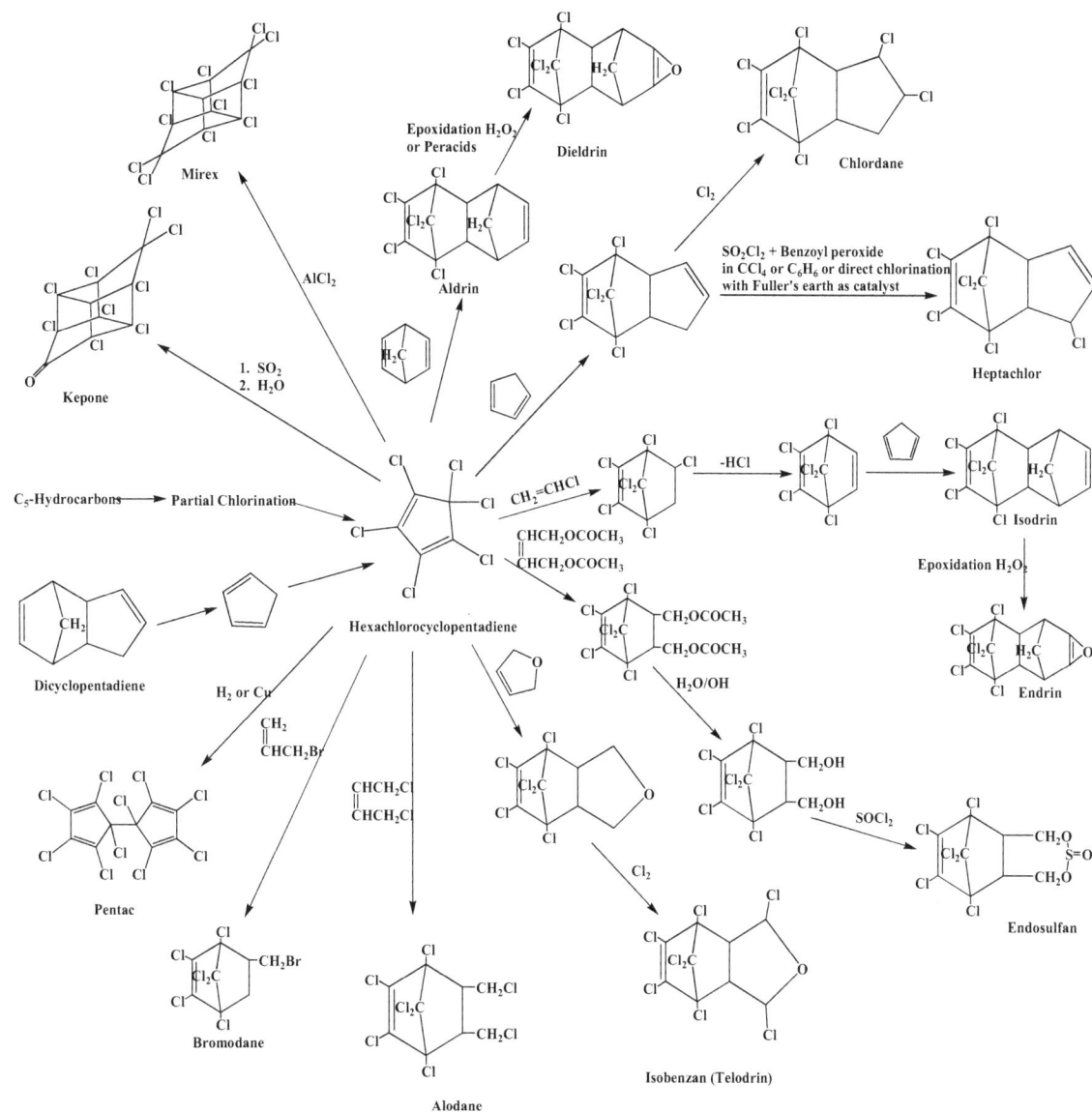

**Figure 3.** Synthesis of chlorinated Cyclopentadiene pesticides from Hexaclorocyclopentadiene.

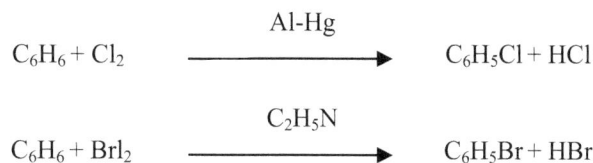

$$C_6H_6 + Cl_2 \xrightarrow{\text{Al-Hg}} C_6H_5Cl + HCl$$

$$C_6H_6 + BrI_2 \xrightarrow{\text{C}_2\text{H}_5\text{N}} C_6H_5Br + HBr$$

The halogen carriers or aromatic halogenation catalysts are usually all electrophilic reagents ($AlCl_3$, $FeCl_3$, $FeBr_3$, and $ZnCl_2$) and their functions appears to increase the electrophilic activity of the halogens.

**DDT:** It is formed by condensation of chloral (1 mol) and monochlorobenzene (2 mol) in the presence of sulphuric acid and gives a mixture which contains *op`*-DDT, *pp`*-DDT , *oo`*-DDT and small amount of other derivatives.

### 1.4.2.3. Cycloparaffins

In principle, hexachlorocyclohexane (HCH) can be made in three different ways,

1.   Cyclohexane, itself chlorinated.

2.   Cyclohexane, may undergo both addition and substitution chlorination.

3.   Benzene may add three chlorine molecules.

The first two methods are of theoretical interest. However, the third is well used method for HCH preparation. Reaction can occur between benzene and chlorine to give either substitution products ranging from monochlorobenzene to hexachlorobenzene (in which all the hydrogen atoms are replaced by the chlorine), or the addition products, the hexachlorocyclohexane, depending on the conditions employed.

### 1.4.2.4. Chlorinated Terpins

Polychlorinated terpins (toxaphene) produced by chlorinating various naturally occurring terpins. Basically, these materials arise by the chlorination of α – pinene or camphene, and the insecticidal properties of the products increase with chlorine content (Fig. **4**).

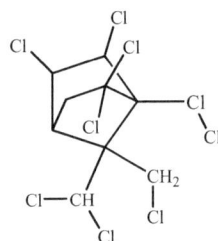

Toxaphene

IUPAC : 2,5-cndo,6-exo,8,9,9,10,10-octachlorobornene-2

FORMULA : $C_{10}H_{10}Cl_8$

**Figure 4.** Structure of Toxaphene.

### 1.4.3. Organophosphorous Pesticides (OP)

Organophosphorous pesticides (often referred as organophosphates) are neutral ester or amide derivatives of phosphorous acids carrying a phosphoryl (P-O) or thiophosphoryl (P-S) group. Certain fluorides and chlorides are also used to develop organophosphorus pesticides, however only one phosphoric acid is known for its insecticidal properties. Gerhard Schrader and his co-workers of Germany, who discovered the insecticidal properties of the first OP pesticides in 1937, insecticide was named after him as Schradan with the general formula,

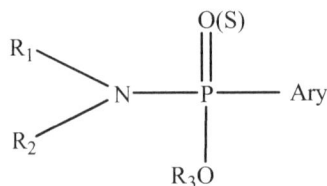

$R_1$, $R_2$, and $R_3$ are alkyl groups, and Ary is inorganic or organic radical (Cl, F, SCN, $CH_3COO$).

OP pesticides are identified by single characteristic that they act by inhibiting cholinesterase enzyme. They are manufactured at very high temperatures (150-200°C), thus they commonly contain isomers or bi-products which are responsible for their unpleasant odour, and for their anti-cholinesterase activities. They are readily activated and degraded in mammals and by micro-organisms and therefore do not accumulate in the environment thus OP pesticides are non-persistent and are reasonably biodegradable. It is because of this feature the OP pesticides have mainly replaced the persistent organochlorine pesticides and become one of the largest group of the pesticides used today. Among the products, parathion is highly toxic to mammals ($LD_{50}$ rat <5 mg/kg body weight), whereas pirimiphos-methyl is less toxic ($LD_{50}$ rat 2000 mg/kg), and are widely used in agriculture.

---

**Learning the Facts:**

Organophosphate pesticides are acute toxicants of high level activity and are non-persistence and biodegradable. The toxicity of OP is due to their ability to inhibit the vital enzyme cholinesterase. Symptoms of poisoning in insect follows the general pattern of nerve poisoning *i.e.* restlessness, hyper excitability, tremors, convulsions, paralysis and death. Death occurred due to maximum cholinesterase inhibition.

---

### 1.4.3.1. Group of Organophosphorus Pesticides

Member of this group possess different physiological activities and they have different vapour tension at room temperature and different solubility in water. They also differ considerably in chemical stability and toxicity to mammals.

On the basis of their structure, organophosphorus pesticides are classified into the following groups, other combinations are also possible.

**1. Orthophosphates:** Chlorpheniphos, Dichlorovos, Mevinphos, Phosphamidon.

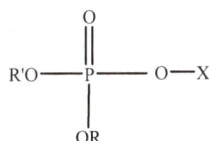

$$R'O-\overset{\overset{\displaystyle O}{\|}}{\underset{\underset{\displaystyle OR}{|}}{P}}-O-X$$

**2. Phosphorothionates**: Bromophos, Diazinon, Femitrothion, Parathion.

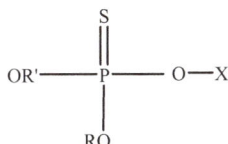

$$OR'-\overset{\overset{\displaystyle S}{\|}}{\underset{\underset{\displaystyle RO}{|}}{P}}-O-X$$

**3. Phosphorothiolates:** Dementon-s-methyl, Vamidothion.

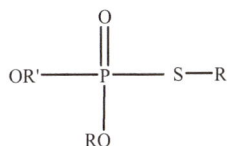

$$OR'-\overset{\overset{\displaystyle O}{\|}}{\underset{\underset{\displaystyle RO}{|}}{P}}-S-R$$

**4. Dithiophosphates or phosphorothiolothionales:** Disulfoton, Phorates, Malathion, Menazon.

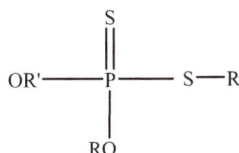

$$OR'-\overset{\overset{\displaystyle S}{\|}}{\underset{\underset{\displaystyle RO}{|}}{P}}-S-R$$

**5. Phosphonates:** Butanoate, Trichlorphon.

$$OR' \longrightarrow \underset{\underset{RO}{|}}{\overset{\overset{O}{\|}}{P}} \longrightarrow X$$

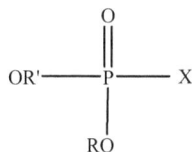

**6. Pyrophosphoramides:** Pyrophosphate.

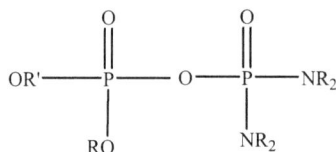

$$OR' \longrightarrow \underset{\underset{RO}{|}}{\overset{\overset{O}{\|}}{P}} \longrightarrow O \longrightarrow \underset{\underset{NR_2}{|}}{\overset{\overset{O}{\|}}{P}} \longrightarrow NR_2$$

### *1.4.3.2. Group characteristics: Further Explanation of 1.4.3.1.*

1. **Group I**: These compounds have low chemical stability and solubility in water. They are rapidly hydrolysed in water. They are used as contact insecticide (Mevinphos, Tetraethylpyrophosphates (TEPP), and Tetrachlorvenphos).

2. **Group II:** These compounds have moderate to high chemical stability. Usually, they have low solubility in water but more soluble in oil. They are persistent, contact or quasi-systemic (partly systemic and partly contact) in nature (Malathion, Methylparathion, Fenintrothion, Diazinon).

3. **Group III:** These compounds have moderate to high chemical stability. Their oil/water partition coefficient enables them to enter in the plant and to be translocated in them. These are systemic pesticides and have to be activated before reaching to the site of contact (Phorate, Dimethoate, Disulpton, and Formothion).

4. **Group IV:** These compounds have high vapour pressure and low chemical stability and because of these feature they are used as fumigants (Dichlorovos, Sulfotep).

5. **Group V:** These compounds are suitable for formulation as granule for soil application (Chlorfenvinphos, Bromophos).

6. **Group VI:** These compounds are suitable for surface application (Cumaphos, Fenchlorphos).

Organophosphate compounds demonstrate high insecticidal and acaricidal activity (Fig. **5**) many of them have fungicidal, nematicidal or herbicidal activity. They easily degrade into non-toxic metabolite and have low persistence in the environment and do not accumulate in the ecosystem. Few among them are systemic in nature, effective at lower concentration with little or no residue. Due to their high toxicity, these compounds are required in low amount. They possess broad spectrum activities against number of insect pests.

OP pesticides have different names in different countries. Thus rules were adopted by agreement between British Chemical Society and American Chemical Society in 1952 to adopt single name for each compound. OP compounds are named as derivative of their corresponding parent compound (acids or hydrides).

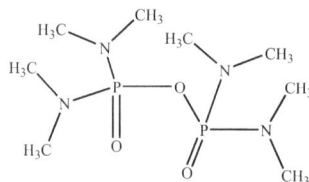

Schradan

IUPAC: octamethylpyrophosphoric tetraamide

FORMULA: $C_8H_{24}N_4O_3P_2$

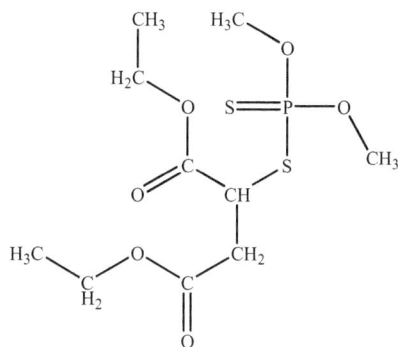

Malathion

IUPAC: diethyl (dimethoxyphosphinothioylthio)succinate or S-1,2-bis(ethoxycarbonyl)ethyl O,O-dimethyl phosphorodithioate

FORMULA: C10H19O6PS2

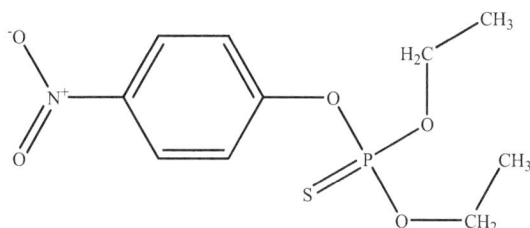

Parathion

IUPAC: O,O-diethyl O-4-nitrophenyl phosphorothioate

FORMULA: $C_{10}H_{14}NO_5PS$

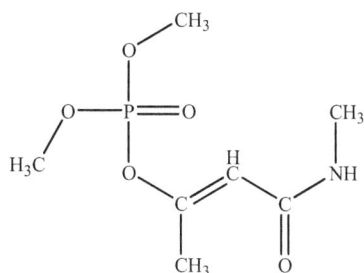

Monocrotophos

IUPAC : dimethyl (E)-1-methyl-2-(methylcarbamoyl)vinyl phosphate or 3-dimethoxyphosphinoyloxy-N-methylisocrotonamide

FORMULA: $C_7H_{14}NO_5P$

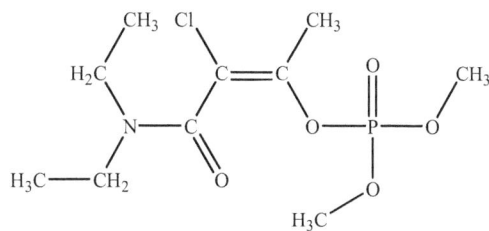

Phosphamidon

FORMULA: $C_{10}H_{19}C_lNO_5P$

IUPAC : (EZ)-2-chloro-2-diethylcarbamoyl-1-methylvinyl dimethyl phosphate or (EZ)-2-chloro-3-dimethoxyphosphinoyloxy-N,N-diethylbut-2-enamide

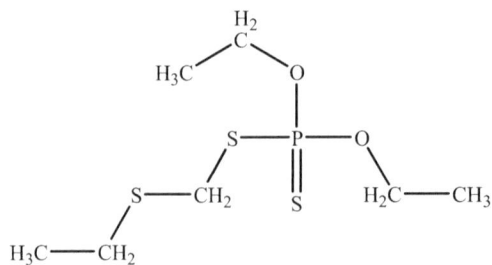

Phorate

IUPAC : O,O-diethyl S-ethylthiomethyl phosphorodithioate

FORMULA: $C_7H_{17}O_2PS_3$

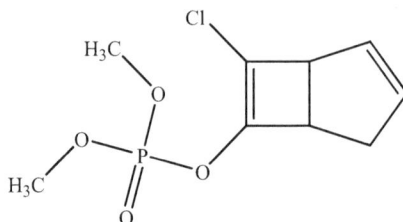

Hepténophos

IUPAC : 7-chlorobicyclo[3.2.0]hepta-2,6-dien-6-yl dimethyl phosphate

FORMULA: $C_9H_{12}C_1O_4P$

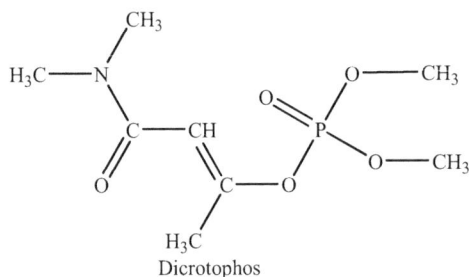

Dicrotophos

IUPAC : (E)-2-dimethylcarbamoyl-1-methylvinyl dimethyl phosphate or 3-dimethoxyphosphinoyloxy-N,N-dimethylisocrotonamide

FORMULA: $C_8H_{16}NO_5P$

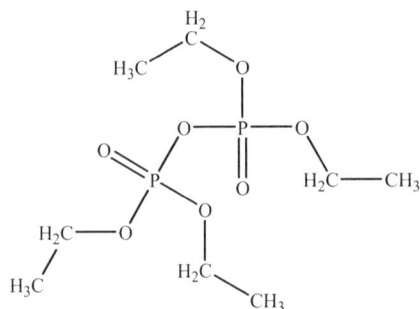

TEEP

IUPAC : Tetraethyl pyrophosphate

FORMULA: $C_8H_{20}O_7P_2$

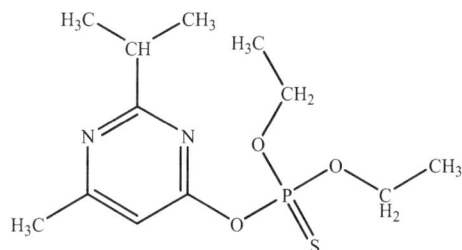

Diazinon

IUPAC : O,O-diethyl O-2-isopropyl-6-methylpyrimidin-4-yl phosphorothioate

FORMULA: $C_{12}H_{21}N_2O_3PS$

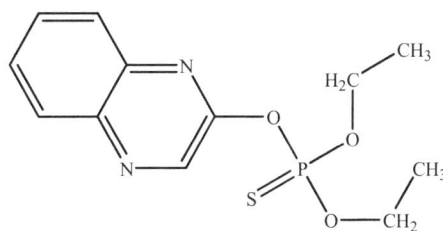

Quinalphos

IUPAC : O,O-diethyl O-quinoxalin-2-yl phosphorothioate

FORMULA: $C_{12}H_{15}N_2O_3PS$

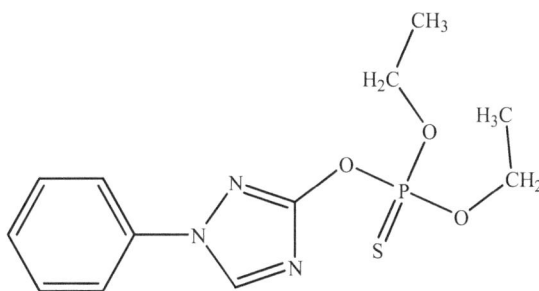

Triazophos

IUPAC : O,O-diethyl O-1-phenyl-1H-1,2,4-triazol-3-yl phosphorothioate

FORMULA: $C_{12}H_{16}N_3O_3PS$

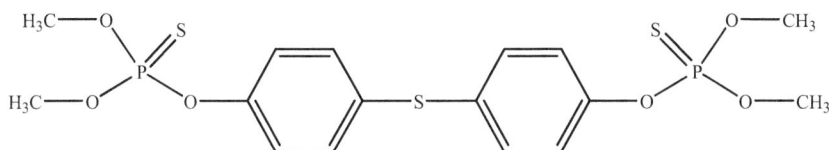

Téméphos

IUPAC : O,O,O',O'-tetramethyl O,O'-thiodi-p-phenylene bis(phosphorothioate) or O,O,O',O'-tetramethyl O,O'-thiodi-p-phenylene diphosphorothioate

FORMULA: $C_{16}H_{20}O_6P_2S_3$

**Figure 5.** Structure of different organophosphorus pesticides.

## 1.4.4. Carbamate Pesticides

Carbamates are anticholinesterase inhibitor pesticides and are synthetic derivative of physostigmine, also known as eserine, which is a principle alkaloid of plant, *Physostigma venenosum,* Calabar bean.

Chemically, they are esters of carbamic acid ($HOOC.NH_2$) with insecticidal (and related) properties possess the general structure,

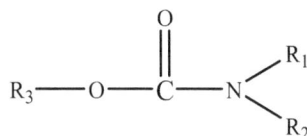

$R_1$, and $R_2$ are hydrogen, methyl, ethyl, propyl or other short chain alkyls and $R_3$ is phenol, napthyl ring or other cyclic hydrocarbons or oxime derivative (Fig. **6**). There are three major groups of carbamates,

1.  **Group I:** Comprises N-methyl carbamate esters of phenols that is the compounds with a hydroxyl group attached directly to a phenyl or napthyl ring *i.e.* Carbaryl (1-napthyl N-methyl carbamate).

2.  **Group II:** Comprises N-methyl and N-dimethyl esters of heterocyclic phenols *i.e.* Carbofuran (2,3-dihydro-2,2-dimethyl benzofuran-7-yl N-methyl carbamate).

3.  **Group III:** Contains oxime (the OH group of which has been carbamylated) *i.e.* Aldicarb.

Their distinctive feature is low toxicity to mammals (exception is aldicarb) and broad spectrum to insect control (used widely for lawn and garden insects). Carbamate pesticides posses both contact and stomach toxicity. They are nerve poisons that inhibit acetyl cholinesterase at nerve synapses, causes rapid twitching of the muscles, in-coordination (ataxia), convulsions, paralysis and death. Carbamates are degraded by many enzymatic catalysed reactions, primarily through hydrolysis, oxidation and conjugation. Their usage are diverse, some of them are used extensively for forest protection, while others are widely used against insect pests of potatoes and grains e.g. Carbaryl (Sevin) and Aldicarb (Temik).

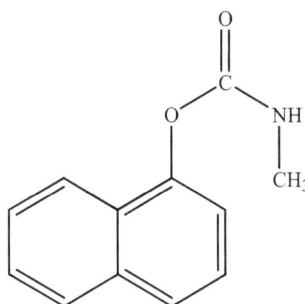

Carbaryl

IUPAC: 1-naphthyl methylcarbamate

FORMULA: $C_{12}H_{11}NO_2$

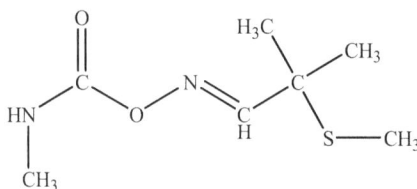

Aldicarb

IUPAC: (EZ)-2-methyl-2-(methylthio)propionaldehyde O-methylcarbamoyloxime

FORMULA: $C_7H_{14}N_2O_2S$

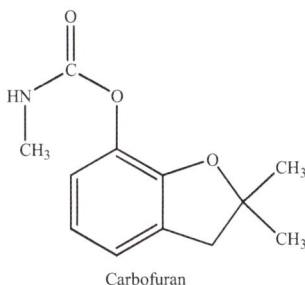

Carbofuran

IUPAC : 2,3-dihydro-2,2-dimethylbenzofuran-7-yl methylcarbamate

FORMULA: $C_{12}H_{15}NO_3$

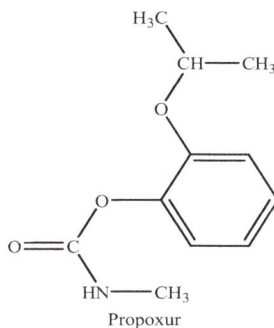

Propoxur

IUPAC: 2-isopropoxyphenyl methylcarbamate

FORMULA: $C_{11}H_{15}NO_3$

**Figure 6.** Structure of different carbamate pesticides.

**Learning the Facts :**

The most important reaction for degradation of carbamate is hydrolysis. N-alkyl group stabilizes the carbamate ester. The order of stability is N-isopropyl < N-ethyl < N-n-propyl < N-methyl and the corresponding N-N-disubstituted carbamates are found to be more resistant to hydrolysis then their mono analogues. Factor like temperature, pH and hydroxyl ion concentration also play a role in carbamate decomposition. The rate of hydrolysis varies directly with increase in temperature while the half-life decreases with the increase of pH value. In acidic solution, the hydrolysis of carbamate generally occurs at very low rate.

### 1.4.5. Pyrethroid pesticides:

Pyrethroids are natural esters formed by combination of two carboxylic acid and three keto acid. It is extracted from plant *chrysanthemum cinerariaefolium,* in which flower contains an average 1.3 percent pyrethrins, it was first used as powder in around 1851. Pyrethrum concentrate was prepared from flower by extracting with petroleum ether or methanol or acetone or dichloromethane or ethylene dichloride. Technical pyrethrum contains 20-30 percent toxic ingredient. Oral $LD_{50}$ is around 1500 mg/kg body weight for rat, but very toxic for insects. The pyrethrins are esters formed by a combination of two acids *i.e.* chrysanthemic acid and pyrethric acid, with three alcohols (rethrolones), these are pyrethrolone, cinerolone and jasmolone. There are four principal active ingredients in pyrethrum flowers, known as *pyrethrins I and II* and *cinerins I and II*. All four are esters comprising an acid (Fig. 7). In addition, small quantities of *jasmolins I and II* are present, these differ from the pyrethrins only in that one double bond in the side chain of the alcohol moiety of pyrethrins is saturated. The acid present in compounds designated by *I* is called chrysanthemic acid while in those designated by *II* is called pyrethric acid. Although natural source pyrethrum has been used for hundreds of years. *Synthetic Pyrethroid (SP)* group of insecticides was introduced more recently, in the early 1970s. SPs are structural analogues of the natural pyrethrums (Fig. **8**). They are more stable to light and possess a higher insecticidal activity, almost ten times that of most organophosphates and carbamate insecticides. They are nerve poisons and affect the nerve axon, causing repetitive discharge of nerves which results in eventual paralysis. The stability and activity of the synthetic pyrethroids are reflected in their increased use during the last two decades on fruits, vegetables and corn. The high insecticidal activities of these chemicals allow relatively small amounts to be applied. They also have good biodegradability due to the ester linkages. Thus because of these two reasons their environmental residues are almost absent. Their principle disadvantage is the very broad insecticidal activity which tends to eliminate many beneficial. Also all pyrethroids are lipophillic in nature, and practically insoluble in water e.g. Allethrin and Permethrin.

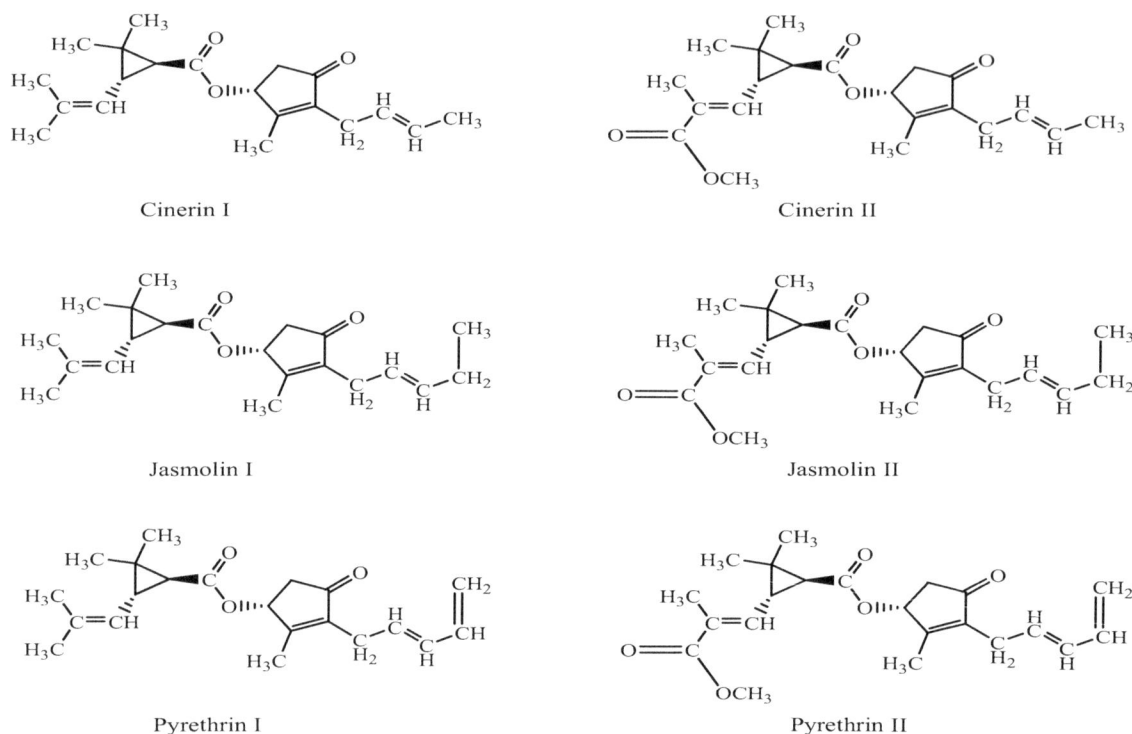

Cinerin I

Cinerin II

Jasmolin I

Jasmolin II

Pyrethrin I

Pyrethrin II

Pyréthrines

**Figure 7.** Constituent of pyrethrins.

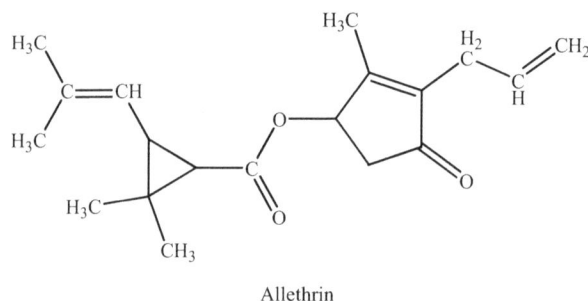

Allethrin

IUPAC : (*RS*)-3-allyl-2-methyl-4-oxocyclopent-2-enyl (1*RS*,3*RS*;1*RS*,3*SR*)-2,2-dimethyl-3-(2-methylprop-1-enyl)cyclopropanecarboxylate or (*RS*)-3-allyl-2-methyl-4-oxocyclopent-2-enyl (1*RS*)-*cis-trans*-2,2-dimethyl-3-(2-methylprop-1-enyl)cyclopropanecarboxylate or (±)-3-allyl-2-methyl-4-oxocyclopent-2-enyl (±)-*cis-trans*-chrysanthemate

FORMULA : $C_{19}H_{26}O_3$

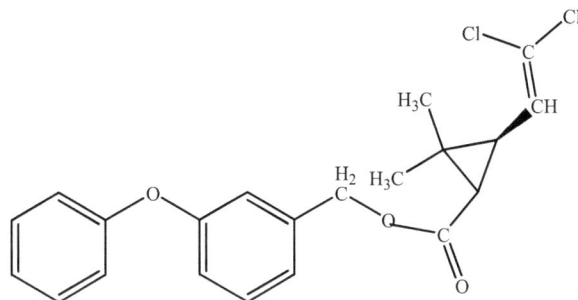

Peremethrin

IUPAC : 3-phenoxybenzyl (1*RS*,3*RS*;1*RS*,3*SR*)-3-(2,2-dichlorovinyl)-2,2-dimethylcyclopropanecarboxylate or 3-phenoxybenzyl (1*RS*)-*cis-trans*-3-(2,2-dichlorovinyl)-2,2- dimethylcyclopropanecarboxylate

FORMULA : $C_{21}H_{20}Cl_2O_3$

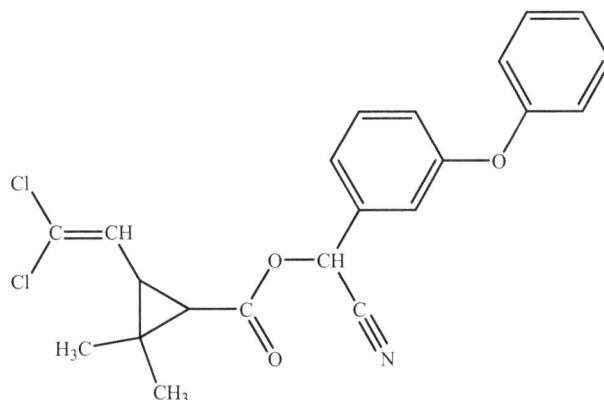

Cypermethrin

IUPAC : (*RS*)-α-cyano-3-phenoxybenzyl (1*RS*,3*RS*;1*RS*,3*SR*)-3-(2,2-dichlorovinyl)-2,2-dimethylcyclopropanecarboxylate or (*RS*)-α-cyano-3-phenoxybenzyl (1*RS*)-*cis-trans*-3-(2,2-dichlorovinyl)-2,2-dimethylcyclopropanecarboxylate

FORMULA: $C_{22}H_{19}Cl_2NO_3$

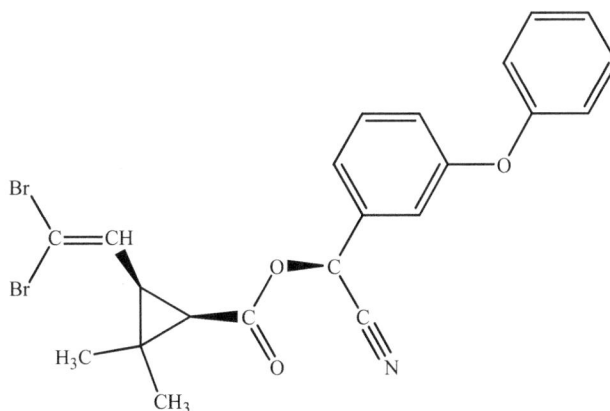

Deltamethrin

IUPAC: (*S*)-α-cyano-3-phenoxybenzyl (1*R*, 3R)-3-(2, 2-dibromovinyl)-2, 2-dimethylcyclopropanecarboxylate or (*S*)-α-cyano-3-phenoxybenzyl (1*R*)-*cis*-3-(2, 2-dibromovinyl)-2, 2-dimethylcyclopropanecarboxylate

FORMULA: $C_{22}H_{19}Br_2NO_3$

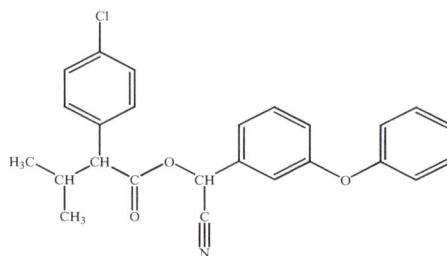

Fenvalérate

IUPAC : (*RS*)-α-cyano-3-phenoxybenzyl (*RS*)-2-(4-chlorophenyl)-3-methylbutyrate

FORMULA: $C_{25}H_{22}ClNO_3$

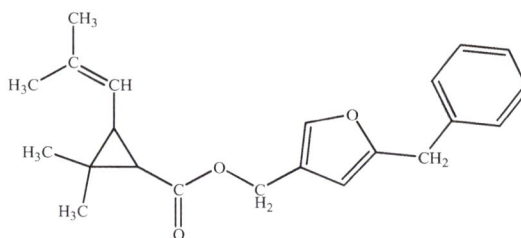

Resméthrine

IUPAC: 5-benzyl-3-furylmethyl (1RS, 3RS; 1RS, 3SR)-2, 2-dimethyl-3-(2-methylprop-1-enyl) cyclopropanecarboxylate

FORMULA: $C_{22}H_{26}O_3$

**Figure 8.** Structure of Pyrethrins.

**Learning the Facts :**

Pyrethroids are esters formed by the combination of two acids i.e Chrysenthemic acid and Pyrethric acid, with three alcohols namely Pyrethrolone, Cinerolone and Jasmolone. Crysanthemic acid is produced industrially in a cyclopropanation reaction of a diene as a mixture of cis- and trans- isomers followed by hydrolysis of the ester. The ester of Crysenthemic acid are pyrethrin-I, cinerin-I and jasmolin-I and are together known as Pyrethrin-I. The esters of pyrethric acid are pyrethrin-II, cinerin-II and jasmolin-II and are together known as Pyrethrin-II. pyrethrin-I and II, cinerin-I and II comprising of an acid containing a 3C ring joined to an alcohol containing a 5C ring while jasmolins I and II are differ from the pyrethrins only in that one double bond in the side chain of the alcohol moiety of pyrethrins is saturated.

## 1.4.6. Plant Origin Pesticides

Plants are known to produce a diverse range of secondary metabolites such as alkaloids, flavonoids, polyacetylenes, terpenoids, *etc.* Many of these chemicals protect the plant from pests and pathogens. More than 2400 plant species belonging to 235 families have been reported to possess pest control properties. Flowers, leaves and roots are finely grounded and used, or toxic ingredients of plants are extracted and used alone or in mixture. This method of insect control has been in use for centuries. Advantages of using botanicals are that they are safe to natural enemies, and being biodegradable does not leave toxic residues.

## 1.4.7. Nicotine

The main sources of nicotine are the two species *Nicotiana tabacum* and *N. rustica,* the latter being more abundant in India. Free nicotine is a colourless or pale-yellow oily liquid. It has an odour of pyridine, because of its high volatility free nicotine is mainly used as a fumigant (Fig. **9**). In agriculture nicotine is used as nicotine sulphate which acts as a stomach poison. Addition of alkaline compounds such as soap and calcium caseinate at the time of spraying liberates the nicotine more quickly, making it a more effective contact insecticide or fumigant. Oral $LD_{50}$ of nicotine sulphate to rat is 83 mg/ kg body weight and dermal 285 mg/ kg body weight. Nicotine is a nerve poison and mimics acetylcholine at the nerve synapse.

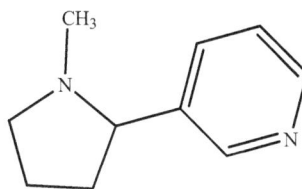

Nicotine

IUPAC: 3-[(2*S*)-1-methylpyrrolidin-2-yl] pyridine

FORMULA: $C_{10}H_{14}N_2$

**Figure 9.** Structure of Nicotine.

### 1.4.8. Rotenone

It is mainly obtained from the roots of two species of Derris which grow in Far East and some species of *lonchocarpus* which grow in Amazon valley in South America. Natives throughout the tropics have used rotenone-containing plants as fish poisons. Mode of action of rotenone poisoning derives from the ability of rotenone to inhibit the respiratory metabolism or the electron transport system between the NADH dehydrogenases and the coenzyme Q at complex I. Rotenone is both a contact as well as a stomach poison (Fig. **10**). It has very low toxicity to mammals, and thus is particularly useful in killing external parasites of livestock such as lice, fleas and ticks in dust form. It can be used as a lotion for chiggers, as an emulsion for scabies, and as a spray against cattle grubs and mange for dogs. The only drawback associated with the use of rotenone as an effective botanical is that it deteriorates in storage and has slow action against some insects.

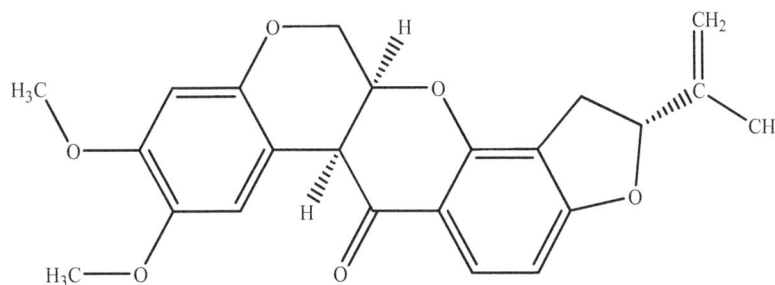

Rotenone

IUPAC: (2*R*,6a*S*,12a*S*)-1,2,6,6a,12,12a-hexahydro-2-isopropenyl-8,9-dimethoxychromeno[3,4-*b*]furo[2,3-*h*]chromen-6-one

FORMULA: $C_{23}H_{22}O_6$

**Figure 10.** Structure of Rotenone

### 1.4.9. Others

Neem (oil extracts of neem seed kernel, Fig. **11**), Chinaberry (a close relative of Neem tree), Pongram, Custard apple (powdered seeds of custard apple), Ryania (roots of shrub), Limonene (citrus peels), Sabadilla (seeds of lily).

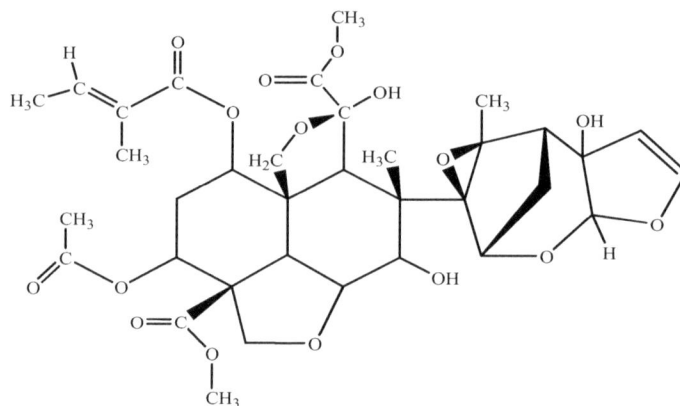

Azadirachtin

IUPAC : dimethyl (2a*R*,3*S*,4*S*,4a*R*,5*S*,7a*S*,8*S*,10*R*,10a*S*,10b*R*)-10-acetoxy-3,5-dihydroxy-4-[(1a*R*,2*S*,3a*S*,6a*S*,7*S*,7a*S*)-6a-hydroxy-7a-methyl-3a,6a,7,7a-tetrahydro-2,7-methanofuro[2,3-*b*]oxireno[*e*]oxepin-1a(2*H*)-yl]-4-methyl-8-{[(2*E*)-2-methylbut-2-enoyl]oxy}octahydro-1*H*-naphtho[1,8a-*c*:4,5-*b'c'*]difuran-5,10a(8*H*)-dicarboxylate

FORMULA: $C_{35}H_{44}O_{16}$

**Figure 11.** Structure of Azadirachtin

## 1.5. BIOPESTICIDES

Biopesticides are natural biological control agents used either by introducing new species into the environment or by increasing the effectiveness of those already present. Traditionally, this method was employed to control insect pests by parasitoids, predators and pathogens. A parasite is an organism which at one time or other lives in the body of the host and may or may not kill the host, after the completion of its development. A parasitoid is an organism which completes its life cycle on the host and then kills it. A predator on the other hand is a free living animal and kills its prey immediately. In 1949, E.A. Steinhaus had coined another term *'microbial control'*. It employs those microorganisms or their products that are capable of attacking or killing insect pests. The advantage of the use of microorganisms for insect pest control are (i) they have minimum effect on non-target organism as they are highly host specific and (ii) these microorganisms have a natural capability of causing diseases at epizootic levels due to their persistence in soil and their efficient transmission. Drawbacks associated with its use are (i) microbial pesticides are not economically viable and (ii) they are relatively slow in action. Bioinsecticide usage involves three major techniques viz. introduction, conservation and augmentation.

1.  **Introduction:** It involves the release of biopesticides when the pest is to be controlled and during the vulnerable stage of insect development.

2.  **Conservation:** It means to avoid the destruction of natural enemies and the use of measures that increase their longevity, reproduction and attractiveness of an area to natural enemies.

3.  **Augmentation:** It includes all activities designed to increase numbers or effect of existing natural enemies. These objectives may be achieved by releasing additional number of a natural enemy into a system or modifying the system in such a way as to promote greater numbers or effectiveness. The releases have to be made periodically. This may be done in two ways,

    a)  *Inoculative releases:* Released once in a year. The purpose of it is to re-establish a species of natural enemy which is otherwise less in number due to unfavourable conditions. In this case, the control is expected from the progeny and subsequent generations, and not by the release itself.

b) *Inundative releases:* Involves mass culture and release of natural enemies to suppress the pest population directly. These are most economical against pests with one or utmost few discrete generations every year.

## 1.6. NEONICOTINOIDS AND NITROGENOUS PESTICIDES

The neonicotinoids are the new class of insecticides, derived synthetically from nicotinoids (Fig. **12**). It includes acetamiprid, clothianidin, dinotefuran, imidacloprid, nitenpyram, thiacloprid, and thiamethoxam. The biotransformation of these compounds involves activation reactions but largely it is detoxification mechanism. It is opposite to nicotine, epibatidine, and other ammonium nicotinoids, which are mostly protonated at physiological pH. The neonicotinoids are not protonated and have electronegative nitro or cyano pharmacophore. These substitutions increase the hydrophobicity of the compound. Thus, the neonicotinoids are also works as systemic poisons. Neonicotinoids acts as agonists at the insect nicotinic acetylcholine receptor (nAChR), whereas the nicotine acts at the same target. Fundamental differences between the nAChRs of insects and mammals confer remarkable selectivity for the neonicotinoids. Nicotinoids, with no ionisation have poor binding affinity with the insect nAChR. The presence of electron donating atom increases the binding affinity with the insect nAChR. Thus the neonicotinoids have higher toxicity to insects in comparison to nicotinoids. In vertebrates nAChR prefers the presence of one unit of positive charge for stronger binding. The ionized nicotine binds at an anionic sub-site in the mammalian nAChR. Thus resulting in high toxicity of nicotinoids for mammals.

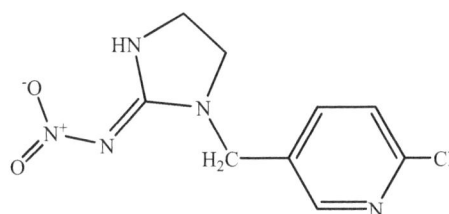

Imidacloprid

IUPAC: (*E*)-1-(6-chloro-3-pyridylmethyl)-*N*-nitroimidazolidin-2-ylideneamine

FORMULA: $C_9H_{10}ClN_5O_2$

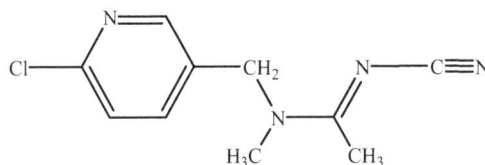

Acetamiprid

IUPAC: (*E*)-*N$^1$*-[(6-chloro-3-pyridyl) methyl]-*N$^2$*-cyano-*N$^1$*-methylacetamidine

FORMULA: $C_{10}H_{11}ClN_4$

**Figure 12.** Structure of Neonicotinoids.

## 1.7. ACRONYMS

**Terms**

**Fumigants**

Fumigants are insecticides in the form of gases and are slightly heavier than air and have the ability to spread to all sealed areas (Fig. **13**). The toxicity of a fumigant depends on the respiration rate of the target organism. Generally, lower the temperature, lower is the respiration rate of the organism which tends to make the pest less susceptible. Fumigation at lower temperatures thus requires a higher dosage and longer exposure period. Fumigants differ significantly in their mode of action. Some kill the pest rapidly while for others their action is slow. In sub lethal dosages, some fumigants may paralyze the pest allow the pest to recover. Fumigants usually have no effect on fumigated commodities but some of them have detrimental effects even at low concentration.

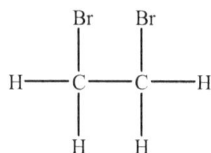

Ethylene dibromide

IUPAC: ethylene dibromide or 1, 2-dibromoethane

FORMULA: $C_2H_4Br_2$

**Figure 13.** Structure of Ethylene dibromide.

## REFERENCES

[1]  The pesticide Manual. Clive Tomlin, Ed., Crop Protection Publication, UK, 1997.
[2]  Agrochemicals. Franz Muller, Ed., WILEY-VCH, Federal Republic of Germany, 2000.
[3]  Pesticide Biochemistry and Physiology. C.F. Wilkinson, Ed., Plenum Press, NY, 1976.
[4]  Dileep K. Singh, Toxicology of insecticides. http://nsdl.niscair.res.in/ CSIR, New Delhi, India, 2007.
[5]  Insecticides in Agriculture and Environment. A.S.Perry, I.Yamamoto, I. Ishaaya and R.Perry. Narosa Publishing House, New Delhi, India, 1998.
[6]  The Merck Index 13th Edition, 2001.
[7]  Pesticide Development: A Brief look at the History. A. Gordon Holley, Melanie Kirk (Eds.). Southern Regional Extension Forestry, 2007.

## QUESTIONS

1.  Describe the following terms in 100 words.

    (a) Insecticide (b) Acaricide, (c) Nerve poisons (d) Repellents (e) Fumigants.

2.  Differentiate between,

    (a) Fungicide and weedicide, (b) Insecticide and herbicide (c ) Biopesticides and botanical pesticides

3.  Explain the group characteristics of the organophosphorus pesticides. Why they are non persistant in the environment?

4.  Explain briefly, the group characteristics of organochlorine pesticides. Why they are persistant in the environment?

5.  Write a short notes on the followings

    (a) Fumigants.

    (b) Acetyl cholinesterase.

    (c) Neonicotenoids and nitrogenous pesticides.

    (d) Neonicotinoids

# CHAPTER 2

## Individual Pesticides

**Abstract:** In this chapter, individual representative pesticides are described, taking as example DDT, DDD, HCH, γ-Lindane, Endosulfan, Malathion, Parathion, Chlorpyrifos, Carbaryl, Carbofuran, Aldicarb, Cypermethrin, Allethrin, Diflubenzuron, Teflubenzuron, Fenoxycarb and Imidacloprid. Here, the pesticides are described by using their chemical structure, molecular weight, physicochemical properties and fate in the environment.

**Keywords:** DDT; BHC/HCH; Lindane; Endosulfan; Malathion; Parathion; Chlorpyrifos; Carbaryl; Aldicarb; Carbofuran; Allethrin; Cypermethrin; Diflubenzuron; Teflubenzuron; Fenoxycarb; Hydroprene; Bt toxin; Limonin; Nicotine, Rotenone, Azadirachtin; Imidacloprid; Insecticides; Pesticides.

## INTRODUCTION

Pesticides can be grouped into families and further placed as individual molecule belongs to organochlorines, organophosphates, carbamates, pyrethroids, neonicotinoids, botanical and biopesticides. Organochlorine pesticides disrupt the sodium/potassium channel of the nerve system and force the nerves for continuous transmition resulting into hyperactivity and tremors. Organophosphate and carbamates both inhibits the enzyme acetylcholinesterase and transfer the nerve impulses continuously again resulting into paralysis. Other pesticides have their individual mode of action.

### 2.1. ORGANOCHLORINE PESTICIDE

### 2.1.1. DDT: 2, 2-bis-(p-Chlorodipheny)-1, 1, 1-Trichloroethane

An organochlorine insecticide, prepared by reacting chloral (or its alcholate or hydrate) with chlorobenzene in presence of sulphuric acid, oleum or chlorosulfonic acid. The insecticidal properties of DDT were discovered by Paul Muller of J.R.Geigy, A.G. in Switzerland in 1939 (Fig. **1**). DDT and its metabolites accumulate in body fat and other tissues, either as DDT, DDD or DDE. Under normal circumstances in body a platue level is reached where intake and storage are in equilibrium with excretion, therefore the amount stored in fat will remain constant. Its chemical structure is,

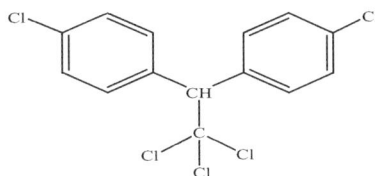

**Figure 1.** Structure of p, p'- DDT.

- **Trade or other names**: DDT, Anofex, Gyron, Cesarex, Chlorophenothane, Guesapon, Guesarol, Gexarex.

- **Appearance:** Technical product p, p'-DDT is white tasteless, almost odourless crystalline solid.

- **Molecular formula:** $C_{14}H_9Cl_5$

- **Molecular Weight:** 354.49

- **Solubility:** Insoluble in water, readily soluble in aromatic and chlorinated solvents, moderately soluble in polar organic solvents and petroleum oils. Solubility in acetone (58 gm/100 ml), cyclohexane (116 gm/100 ml), benzene (106 gm/100 ml), carbontetrachloride (45 gm/100 ml), ethylether (28 gm/100 ml), petroliumether (4-10 gm/100 ml),ethanol (2 gm/100ml) and water (0.0012 ppm) .

**Dileep K. Singh**
**All rights reserved - © 2012 Bentham Science Publishers**

- **Melting Point:** 108.5-109°C

- **Vapour pressure:** 1.5 x $10^{-7}$ mm Hg at 20°C

- **ADI:** 0.005 mg/kg/b.w./d (man)

- **Technical grade:** DDT is actually a mixture of three isomers, predominantly the p, p'-DDT is the main isomer (85%), with the o, p'-DDT and o, o'-DDT isomers present in smaller amounts.

- **Toxicity, single dose:** Rat (male): Oral $LD_{50}$ is 250 mg/kg, Dermal 250-500 mg/kg in oil, 3000 mg/kg as powder. Rat (female): Dermal 2510 mg/kg as powder.

- **Common formulations:** Wettable powders, dusts, aerosols, smoke candles, E.C., *etc.* In house hold formulation DDT is combined with synergized pyrethrins. Concentrations of solid and liquid formulations varied mostly between 20 to 25%.

- **Mode of action:** Central nervous system stimulant producing hyperactivity and tremor, convulsions may occur but are less common than any other organochlorine pesticides.

- **Breakdown in Soil and Groundwater:** DDT is highly persistent in the environment, with a reported half life between 2 to15 years and is immobile in most soils. Routes of loss and degradation include runoff, volatilization, photolysis and biodegradation by aerobic and anaerobic microbes.

## 2.1.2. BHC/HCH: 1, 2, 3, 4, 5, 6-Hexachlorocyclohexane

BHC (benzene hexachloride) was first prepared in 1825 by Michael Faraday, who did not recognize its insecticidal properties. It is produced by the chlorination of benzene under u.v. light. Insecticidal properties are present in γ - isomer. Its chemical structure is (Fig. **2**),

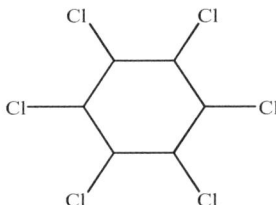

**Figure 2.** Structure of BHC/HCH.

- **Trade or other names**: Benhexachlor, BHC, HCH, Hexachloran, Hexachlor.

- **Appearance:** Technical HCH is off-white to brown, amorphous powder with a characteristic musty odour.

- **Molecular formula:** $C_6H_6Cl_6$

- **Molecular Weight:** 290.80

- **Water Solubility:** 7.3 mg/L at 25 °C.

- **Melting Point:** Crude BHC/HCH begins to melt at 65°C

- **Vapours pressure:** 0.06 mm Hg at 40°C α isomer, 0.17 mm Hg at 40°C β isomer.

- **Technical grade:** Technical HCH consists of various stereoisomers (molecules with a unique structural arrangement, but identical chemical formulas) viz., α isomer constitutes 65-70% (158 °C M.P), β isomer constitutes 5-6% (200 °C M.P), γ isomer constitutes 13% (108-111 °C M.P), δ isomer constitutes 6% (129-130 °C M.P), and ε isomer constitutes 3-4% *etc.*

- **Toxicity, single dose:** Rat: Oral $LD_{50}$ is 500 mg for α isomer/kg, 6000 mg for β isomer/kg, 1000 mg for δ isomer/kg. The β unlike the α and δ isomers, has high chronic and cumulative toxicity.

- **Common formulations:** Emulsifiable concentrate, wettable and dustable powder. The concentration of γ isomer should be stated for compound toxicity.

- **Mode of action:** The α and γ isomers are stimulants of central nervous system, with the principal symptom being convulsions. The β and δ isomers are depressants of the central nervous system.

- **Breakdown in Soil and Groundwater:** BHC/HCH is highly persistent in most soils, with a field half-life of approximately 290 days. When sprayed on the surface, the half-life was shorter than when it is incorporated in the soil. It is very stable in water environment, and is resistant to photo-degradation. It will disappear from the water by secondary mechanisms such as adsorption on sediment, biological breakdown by microbes, and adsorption by fish through gills, skin, and food.

### 2.1.3. Lindane : 1α,2 α,3 β,4α,5α,6β –Hexachlorocyclohexane

In 1912, Van der Linden discovered four isomers of BHC/HCH. Insecticidal properties are present in γ isomer. The pure form of γHCH is named as Lindane in honour of Van der Linden. Lindane is a nerve poison that interferes with GABA neurotransmitter function. It interacts with the $GABA_A$ receptor-chloride channel complex at the picrotoxin binding site. It is produced by the selective crystallization of crude HCH (Fig. **3**).

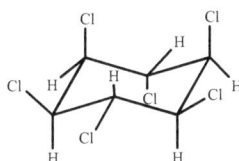

**Figure 3.** Structure of Lindane.

- **Trade or other names**: Lindane, Agrocide, Ambrocide, gamma-hexachlor, gamma benzene hexachloride, Gamaphex, gamma-BHC, Gamma-Col, gamma-HCH, Gammexane

- **Appearance:** Lindane is a colourless crystal compound.

- **Molecular formula:** $C_6H_6Cl_6$

- **Molecular Weight:** 290.80

- **Solubility:** 10 mg/L at 25 °C in water, slightly soluble in petroleum oils, soluble in acetone, aromatic and chlorinated hydrocarbons.

- **Melting Point:** 112.9°C

- **Vapour pressure:** $9.4 \times 10^{-6}$ mm Hg at 20°C

- **Technical grade:** Technical Lindane is comprised of the gamma-isomer of hexachlorocyclohexane.

- **Toxicity, single dose:** Rat: Oral $LD_{50}$ is 88-91 mg/kg, dermal is 900-1000 mg /kg

- **ADI:** 0.008 mg/kg/day

- **Common formulations:** Suspension, emulsifiable concentrate, fumigant, seed treatment, wettable and dustable powder, and ultra-low volume (ULV) liquid.

- **Mode of action:** Lindane is stimulants of central nervous system, with the principal symptom being convulsions.

- **Breakdown in soil and groundwater:** Lindane is highly persistent in most soils, with a field half-life of approximately 15 months. When sprayed on the surface, the half-life was shorter than incorporated into the soil.

### 2.1.4. Endosulfan : 6,7,8,9,10,10-Hexachloro-1,5,5a,6,9,9a-Hexahydro-6,9-Methano-2,4,3-Benzadioxa-thiepin 3-Oxide

Endosulfan is a chlorinated hydrocarbon insecticide and acaricide of the cyclodiene subgroup. It acts as a contact poison to a wide variety of insects and mites. It is formed by Diels-Alder reaction of hexachlorocyclopentadiene with *cis*-butene-1, 4-diol and subsequent reaction of the adduct with thionyl chloride. Technical endosulfan is a 7:3 mixture of stereoisomers *i.e.* α and β –endosulfan (Fig. **4**).

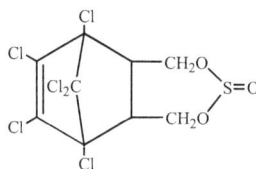

**Figure 4.** Structure of Endosulfan.

- **Trade or other names**: Thiodan, Afidan, Beosit, Cyclodan, Devisulfan, Endocel, Endocide, Endosol, Hexasulfan.

- **Appearance:** Pure endosulfan is a colourless crystal. Technical grade is a yellow-brown in colour.

- **Molecular formula:** $C_9H_6Cl_6O_3S$

- **Molecular Weight: :** 406.96

- **Solubility:** It is practically insoluble in water, moderately soluble in most organic solvents.

- **Melting Point:** Technical material, 70-100°C

- **Vapour pressure:** $1 \times 10^{-5}$ mm Hg at 20°C

- **ADI:** 0.006 mg/kg/b.w/d

- **Technical grade:** Technical endosulfan is a mixture of two isomers viz., α isomer constitutes 70% (106°C M.P), and β isomer constitutes 30% (212°C M.P).

- **Toxicity, single dose:** Rat: Dermal $LD_{50}$ is 30-79 mg/kg.

- **Common formulations:** Emulsifiable concentrate, granules, wettable and dustable powder, and ultra-low volume (ULV) liquid. Combinations are available with other pesticides like dimethoate and parathion-methyl at various concentrations.

- **Breakdown in soil and groundwater:** Endosulfan is moderately persistent in the soil with an average field half-life of 50 days. The two isomers have different degradation times in soil. The half-life for the α- isomer is 35 days and 150 days for the β-isomer under neutral conditions. These two isomers will persist longer under more acidic conditions. The compound degrades in soil by microbes.

## 2.2. ORGANOPHOSPHORUS PESTICIDES

Organophosphate pesticides are synthetic in origin and are normally esters, amides, or thiol derivatives of phosphoric, phosphonic, phosphorothioic, or phosphonothioic acids. Organophosphorous pesticides have following groups,

1. **Pyrophosphates:** Schradan, TEPP.

2. **Dialkylarylphosphate, phosphorothioate, phosphorothionate, phosphorothiolate, and phosphorothiolothioate:** Maximum organophosphorus pesticides belong to this group. They are very toxic and their LD $_{50}$ value varies from 5-55/kg body weight. Metabolites are more toxic than parent compounds. P=O analogs are less stable than P=S analogs such as Parathion.

3. **Phosphorohalides and cyanides:** This group includes mostly nerve gases such as Mipafox.

### 2.2.1. Malathion: *S*-1, 2-bis (Ethoxycarbonyl) ethyl *O, O*-Dimethyl Phosphorodithioate

Malathion is a nonsystemic, wide-spectrum organophosphate insecticide. It was one of the earliest organophosphate insecticides developed and introduced in 1950. Malathion is used for the control of sucking and chewing insects on fruits and vegetables, and is also used to control mosquitoes, flies, household insects, animal parasites (ectoparasites), and head and body lice. Malathion is an insecticide of relatively low human toxicity. However, absorption or ingestion of Malathion into the human body results in its metabolism to malaoxon, which is more toxic than the parent compound (Fig. 5).

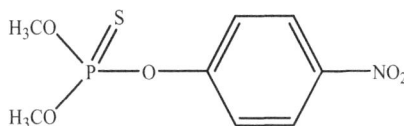

**Figure 5.** Structure of Malathion.

- **Trade or other names:** Malathion is also known as Carbophos, Maldison and Mercaptothion.

- **Appearance:** Technical Malathion is a clear, amber liquid at room temperature.

- **Molecular formula:** $C_{10}H_{19}O_6PS_2$

- **Molecular Weight:** 330.3

- **Solubility:** 145 mg/L at 25 °C in water, miscible with most organic solvents.

- **Melting Point:** 2.85°C

- **Vapour pressure:** $4 \times 10^{-5}$ mm Hg at 30°C

- **ADI:** 0.02 mg/kg/b.w/d

- **Technical grade:** The Technical grade is 95% pure.

- **Toxicity, single dose:** Rat: Oral $LD_{50}$ is 2800 mg/kg.

- **Common formulations:** Suspension, emulsifiable concentrate, wettable and dustable powder, and ultra-low volume (ULV) liquid. Malathion may also be found in formulations with many other pesticides.

- **Breakdown in soil and groundwater:** Malathion is of low persistence in soil with reported field half-lives of 1 to 25 days. Degradation in soil is rapid and related to the degree of soil binding.

### 2.2.2. Parathion: *O, O*-Diethyl *O*-4-Nitrophenyl Phosphorothioate

Parathion is a broad spectrum, organophosphate pesticide used to control many insects and mites (Fig. **6**). It has non-systemic, contact, stomach and fumigant actions. It is produced by the condensation of *O, O*-diethyl phosphorochloridothioate with sodium 4-nitrophenoxide. Parathion is a cholinesterase inhibitor and it acts on the nervous system by inhibiting the acetylcholinesterase.

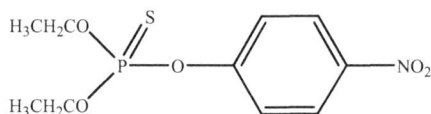

**Figure 6.** Structure of Ethyl Parathion.

- **Trade or other names**: Alkron, Alleron, Aphamite, Corothion, E-605, Ethyl parathion.

- **Appearance:** Pure parathion is a pale yellow liquid with a faint odor of garlic at temperatures above 6°C. Technical parathion is a deep brown to yellow liquid.

- **Molecular formula:** $C_{10}H_{14}NO_5PS$

- **Molecular Weight: :** 291.3

- **Solubility:** 24 mg/L at 25°C in water, slightly soluble in petroleum oils, miscible with most organic solvents.

- **Boiling Point:** 157-162°C /0.6 mm Hg

- **Vapour pressure:** 3.78 x $10^{-5}$ mm Hg at 20°C

- **ADI:** 0.006 mg/kg/b.w/d

- **Technical grade:** Technical grade is 96-98 % pure.

- **Toxicity, single dose:** Rat (male): Oral $LD_{50}$ is 13 mg/kg; Dermal is 21 mg/kg, Rat (female): Oral $LD_{50}$ is 3.6 mg/kg; Dermal 6.8 is mg/kg.

- **Common formulations:** Emulsifiable concentrate, granules, wettable and dustable powder, smokes and aerosol concentrates.

- **Breakdown of Chemical in Soil and Groundwater:** Parathion has little or no potential for groundwater contamination. It binds tightly to soil particles and is degraded by biological and chemical processes within several weeks. Degradation is faster in flooded soil. Photo-degradation may occur on soil surfaces. Sunlight can convert parathion into the active metabolite paraoxon, which is more toxic than parathion.

## 2.2.3. Chlorpyrifos: O, O-Diethyl O-3, 5, 6-Trichloro-2-Pyridyl Phosphorothioate

Chlorpyrifos belongs to phosphorothioate group of organophosphorus pesticides (Fig. 7). It is a broad spectrum insecticide, which affects the ACh esterase receptors on the postsynaptic membrane of the insect nervous system, thus acting as a nerve poison. It has enough limits of safety to mammals and other animals like fish and birds. It shows a wide spectrum of biological activity and is used to control insect pests as well as soil dwelling grubs, *rootworms, borers and subterranean termites.*

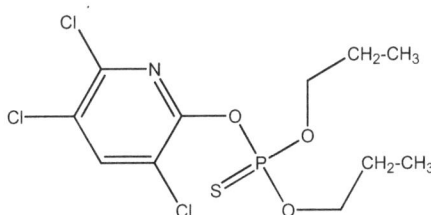

**Figure 7.** Structure of Chlorpyrifos.

- **Trade or other name**: Brodan, Detmol UA, Dowco 179, Dursban, Empire, Eradex, Lorsban, Paqeant, Piridane, Scout, and Stipend.

- **Formulations:** It is available as granules, wettable powder, dustable powder, and emulsifiable concentrate.

- **Appearance:** Technical chlorpyrifos is amber to white crystalline solid with a mild sulfur odor.

- **Molecular Formula :** $C_9H_{11}Cl_3NO_3PS$

- **Molecular Weight:** 350.62

- **Water Solubility:** 2 mg/L at 25 $^0$C .

- **Solubility in organic solvents:** Benzene, Acetone, Chloroform, Carbon disulfide, Diethyl ether, Xylene, Methylene Chloride, Methanol.

- **Melting Point:** 41.5-44 $^0$C

- **Vapour Pressure:** 2.5 mPa 25 $^0$C

- **Toxicity, single dose :** Bird (Chicken) Oral $LD_{50}$ is 30-100 mg/kg

**Breakdown of Chemical in Soil and Groundwater:** Chlorpyrifos is slowly degraded in soil and water with a half life of 60-120 days. In soil and water, the major pathway of degradation of chlorpyrifos is the cleavage of phosphorus ester bond to form 3, 5, 6-trichloro-2-pyridinol (TCP). TCP is further degraded by microbs into carbon dioxide and organic matter.

## 2.3. CARBAMATE PESTICIDES

Carbamates are a group of organic compounds sharing a common functional group with the general structure -NH (CO) O-. The parent compound of all carbamates is called carbamic acid or $NH_2COOH$ and examples are Carbaryl (SEVIN), Oxamyl (VYDATE) Carbofuran (FURADAN) Thiodicarb (LARVIN) Methomyl (LANNATE). They are generally used as insecticides, herbicides and fungicides. Carbamate insecticide inhibits acetylecholinesterase. Herbicide and fungicide carbamate donot inhibit acetylecholinesterase to significant level. Therefore, they are relatively safe or non-toxic to humans.

### 2.3.1. Carbaryl: 1-Napthyl Methylcarbamate

Carbaryl is a contact insecticide of carbamate group with slight systemic properties, produced by the reaction of 1-naphthol with methyl isocyanate or with carbonyl chloride and methylamine. It is primarily used as an insecticide, but can be also used as molluscicide and acaricide (Fig. **8**).

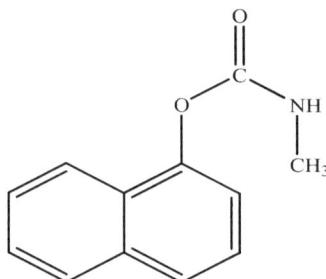

**Figure 8.** Structure of Carbaryl.

- **Trade or other names**: Carbaryl, Adios, Bugmaster, Carbamec, Carbamine, Crunch, Denapon, Dicarbam

- **Appearance:** Carbaryl is a solid that varies from colourless to white or gray, depending on the purity of the compound. But pure Carbaryl is colourless crystalline solid.

- **Molecular formula:** $C_{12}H_{11}NO_2$

- **Molecular Weight: :** 201.2

- **Solubility:** 120 mg/L at $30^{\circ}C$ in water, soluble in most organic solvents such as dimethylformamide and dimethyl sulphoxide.

- **Melting Point:** $142^{\circ}C$

- **Vapor pressure:** $< 4 \times 10^{-5}$ mm Hg at $25^{\circ}C$

- **ADI:** 0.01 mg/kg/b.w/d

- **Technical grade:** Technical product is almost 99% or slightly less pure. It is compatible with most other pesticides except those strongly alkaline, such as Bordeaux mixture or lime sulphur, which hydrolyse it to 1-naphthol.

- **Toxicity, single dose:** Rat (male): Oral $LD_{50}$ is 850 mg/kg, dermal > 4000 mg/kg.

- **Common formulations:** Emulsifiable concentrate, granules, wettable and dustable powder, bait pellets, micronised suspensions in molasses, in non-phytotoxic oil or in aqueous media and as true solutions in organic solvents.

- **Breakdown in soil and groundwater:** Carbaryl has a low persistence in soil. Degradation of Carbaryl in the soil is mostly due to sunlight and bacterial action.

### 2.3.2. Aldicarb : 2-Methyl-2-(Methylthio)Propionaldehyde O Methyl Carbamoyloxime

Aldicarb is a systemic carbamate pesticide used on a variety of crops. It is used to controls insects and nematodes. Aldicarb is a cholinesterase inhibitor, which prevents both the breakdown of acetylcholine and the production of the enzyme cholinesterase in the body and produces the symptoms of neurotoxicity (Fig. **9**).

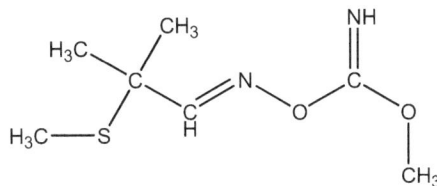

**Figure 9.** Structure of Aldicarb.

- **Appearance:** Aldicarb is a white crystalline solid. It is formulated as a granular mix (10 to 15% active ingredient) because of its toxicity.

- **ADI:** 0.003 mg/kg/day.

- **Molecular Formula** : $C_7H_{14}N_2O_2S$

- **Molecular Weight:** 190.27

- **Water Solubility:** 6000 mg/L at room temperature.

- **Solubility in organic solvents:** Acetone, Xylene, Ethyl ether, Toluene and other organic solvents.

- **Melting Point:** 99-100 $^0$C

- **Vapor Pressure:** 13 mPa at 20 $^0$C

- **Toxicity, single dose:** The oral LD 50 for the female rat is 0.65 mg/kg. Aldicarb is very highly toxic to fish. It is highly toxic to aquatic invertebrates e.g. *Daphnia magna* and freshwater mussels.

- **Breakdown in soil and groundwater:** Aldicarb has negative effects on organisms and pollutes the groundwater. It is highly water soluble and mobile in soils. Aldicarb does not degrade in groundwater. It may easily be carried off into nearby bodies of water through rain and runoff. It is moderately persistent in soil, with half-lives ranging from 9 to 60 days, depending on soil type and conditions.

### 2.3.3. Carbofuran : 2,3-Dihydro-2,2-Dimethylbenzofuran-7-yl Methylcarbamate

Carbofuran is a systemic insecticide and a nematicide and is among the most highly toxic pesticides known to birds (Fig. **10**). It is a cholinesterase inhibitor and most toxic carbamate pesticide. It is used to control the insect pests in potatoes, corn and soybeans. It is effective against a wide range of foliar-feeding and soil pests including nematodes, corn rootworm, rice water weevil, wireworms, sugar-cane borer and others. It is highly toxic to mammals. It has the highest acute toxicity for humans than any insecticide broadly used on field crops.

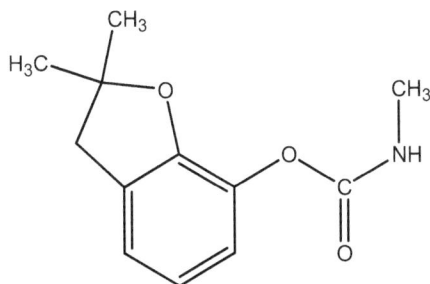

**Figure 10.** Structure of Carbofuran

- **Trade or other name:** Bay 70143, Curater, Furadan, Furacarb, Rampart.

- **Appearance:** Carbofuran is an odorless, white crystalline solid.

- **ADI:** 0.01 mg/kg/day

- **Molecular Formula :** $C_{12}H_{15}NO_3$

- **Molecular Weight:** 221.25

- **Water Solubility:** 320 mg/L at 25 $^0$C

- **Solubility in Other Solvents:** Acetone, Acetonitrile, Benzene, Cyclohexone.

- **Melting Point:** 153-154 $^0$C

- **Vapor Pressure:** 2.7 mPa at 33 $^0$C

- **Molecular Formula :** $C_{12}H_{15}NO_3$

- **Toxicity, single dose:** Carbofuran is highly toxic when ingested or inhalated but moderately toxic when dermal absorption takes place. Carbofuran is highly toxic to birds; $LD_{50}$ for house sparrow is 1.3 mg/kg. Carbofuran is highly toxic to fish. The $LC_{50}$ for bluegill sunfish is 0.24mg/L.

- **Breakdown in soil and groundwater:** Carbofuran is moderately persistent in soils, with half-life ranges from 30 to 120 days. Carbofuran is soluble in water and highly mobile in soils. It has a high potential for groundwater contamination. Carbofuran is degraded by chemical hydrolysis and biodegradation in soil.

## 2.4. PYRETHROIDS

Pyrethrins were originally derived from East African chrysanthemum flowers and shown to have insecticidal activity (Fig. **11**). In the beginning of 1970s, synthetic pyrethroids came into the market for agricultural pest control. They are widely used as home and garden insecticides. They are also used on pets and livestock, mosquito control, treatment of transport vehicles, and for treatment of ectoparasitic disease. Their desirable features are quick knockdown of insects at low application rates and relatively low mammalian toxicity. They are least persistent in the environment. They are effective against a wide range of insect and mite pests and may be mixed with other pesticides for a broad spectrum of pest control usages. Depending on their mode of action pyrethroids are Type I and Type II pyrethroids,

- **Type I pyrethroids:** Natural Pyrethrins, Allethrin, and Resmethrin.

- **Type II pyrethroids :** Cypermethrin ,Deltamethrin, Fenvalerate

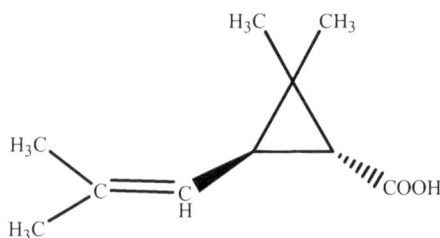

**Figure 11.** Structure of Natural Pyrethrum.

### 2.4.1. Allethrin

(*RS*)-3-allyl-2-methyl-4-oxocyclopent-2-enyl (1*RS*, 3RS; 1RS, 3SR)-2, 2-dimethyl-3-(2-methylprop-1-enyl) cyclopropanecarboxylate

Allethrin is used almost exclusively in homes and gardens for control of flies and mosquitoes, and in combination with other pesticides to control flying or crawling insects (Fig. **12**). It was introduced in 1949 as first synthetic pyrethroid. It is a mixture of several isomeric forms in which the most common form is a 4:1 mixture of the trans- and cis-isomers. D-trans allethrin is usually combined with synergists such as piperonyl-butoxide to increase the toxicity. Allethrin metabolized by a hydrolytic pathway followed by oxidation and dehydrogenation.

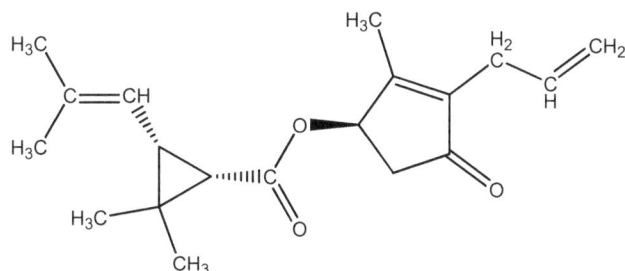

**Figure 12.** Structure of Allethrin.

- **Trade or other names:** Alleviate, Pynamin, D-allethrin, D-cisallethrin, Bioallethrin, Esbiothrin, Pyresin, Pyrexcel, Pyrocide and trans-Allethrin.

- **Appearance:** Allethrin is a clear, amber-colour, viscous liquid with a mild or slightly aromatic odour.

- **Molecular Formula :** $C_{19}H_{26}O_3$

- **Molecular Weight :** 302.4

- **Melting Point:** 4 $^0$C

- **Vapor Pressure:** 16 mPa at 30 $^0$C

- **Solubility:** Allethrin is insoluble in water. Miscible with most organic solvents at 20-25 $^0$C and miscible in petroleum oils. It is soluble in paraffinic and aromatic hydrocarbons.

- **Toxicity, single dose:** Allethrin is slightly to moderately toxic by dermal absorption and ingestion. The oral $LD_{50}$ for allethrin in male rats is 1,100 mg/kg and for female rats is 685 mg/kg. Allethrin is non-toxic to birds. Allethrin is highly toxic to fish and aquatic invertebrates.

### 2.4.2. Cypermethrin

(*RS*)-a-cyano-3-phenoxybenzyl        (1*RS*)-cis,        trans-3-(2,        2-dichlorovinyl)-2,        2-dimethyl-cyclopropanecarboxylate

Cypermethrin is a synthetic pyrethroid of stomach and contact action, produced by the esterification of a-hydroxy-3-phenoxy-phenylacetonitrile with 3-(2, 2-dichlorovinyl)-2, 2-dimethylcyclopropanecarboxylic acid (fi.13). It is a fast-acting neurotoxin for insects. Cypermethrin is highly toxic to fish, bees and aquatic insects.

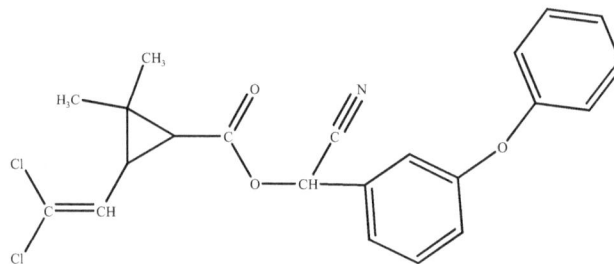

**Figure 13.** Structure of Cypermethrin.

- **Trade or other names**: Ammo, Arrivo, Barricade, Basathrin and Super.

- **Appearance:** Pure isomers of cypermethrin form colourless crystals. When mixed isomers are present, cypermethrin is a viscous semi-solid or a viscous, yellow liquid.

- **Molecular formula:** $C_{22}H_{19}Cl_2NO_3$

- **Molecular Weight:** : 416.3

- **Solubility:** 0.01-0.2 mg/L at 21°C in water, 103g/l in hexane, >450 g/l in acetone at 20°C, soluble in cyclohexane, ethanol, xylene, chloroform.

- **Melting Point:** 60-80 °C (pure isomers).

- **Vapor pressure:** $3.8 \times 10^{-8}$ mm Hg at 70°C (for pure compound).

- **ADI:** 0.05 mg/kg/b.w/d.

- **Technical grade:** Technical cypermethrin is a mixture of eight different isomers, each of which may have its own chemical and biological properties.

- **Toxicity, single dose:** Rat Oral $LD_{50}$ is 303-4123 mg/kg (depending on the carrier and condition used).

- **Common formulations:** It is available as an emulsifiable concentrate or wettable powder.

- **Breakdown in soil and groundwater:** Cypermethrin is moderately persistent in soils. In aerobic conditions, its soil half-life is 4 days to 8 weeks. When applied to a sandy soil under laboratory conditions, its half-life was 2.5 weeks.

## 2.5. INSECT GROWTH REGULATORS

They are synthetic compounds that control insects by disrupting normal growth and development of their larvae nymphs, pupae or adults rather than by toxic action and finally lead to their death. There are three types of insect growth regulators *i.e.* hormonal, enzymatic, and chitin synthesis inhibitors. Growth and molting of immature insects is regulated by three main groups of brain hormones *i.e.* Ecdysone alpha, ecdysone beta, juvenile hormones. Antagonists and analogs of insect growth regulators are also used in pest control. IGR such as juvenile hormones (JH), ecdysones, chitin synthesis inhibitors and other related compounds are used as means for insect growth regulation. Insect growth regulators are eco-friendly in nature. They have low mammalian toxicity and high selective toxicity for insects. Hormonal IGR's may take as long as one generation (3-12 days depending on the insect and the weather condition) to work. Therefore, they are used when populations of pest are low on crop. Their uses are not effective as pest control treatment when outbreaks of the pest are severe.

## 2.5.1. Inhibitors of Chitin Synthesis

### 2.5.1.2. Diflubenzuron: 1-(4-Chlorophenyl)-3-(2, 6-Difluorobenzoyl) Urea

Diflubenzuron is a halogenated benzoylphenyl urea (Fig. **14**). Principle target insect species are the gypsy moth, forest tent caterpillar, several evergreen eating moths and the boll weevil. It is an effective stomach and contact insecticide acting by inhibition of chitin synthesis. All chitin-synthesizing organisms are sensitive to diflubenzuron. Diflubenzuron is the prototypical compound although in this series, second generation compounds also exist.

- **Trade or other names :** Difluron; Dimilin[R]; DU 112307, ENT 29 054, Micromite,

  OMS 1804, PDD6040-I, PH 60-40, TH 6040, Vigilante.

- **Appearance:** Diflubenzuron is a white crystalline solid, the technical material is off-white to yellow crystals.

- **Molecular formula :** $C_{14}H_9ClF_2N_2$

- **Molecular Weight :** 310.7

- **Solubility:** Sparingly soluble in water, 0.2 mg/litre at 20 °C. In polar solvents the solubility is moderate to good depending on the degree of polarity.

- **Toxicity, single dose :** Oral $LD_{50}$ for Rat >4640 mg/kg

- **Breakdown in soil and groundwater:** The rate of degradation in soil is strongly dependent on the particle size of the diflubenzuron and its binding with the soil particles. Their half life varies from a week to month. Almost the entire parent compound breaks down to form DFBA (2, 6-difluorobenzoic acid) and CPU (4-chlorophenyl urea). Under field conditions diflubenzuron has very low mobility.

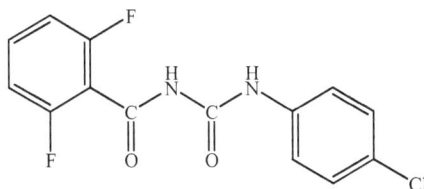

**Figure 14.** Structure of Diflubenzuron.

### 2.5.1.3. Teflubenzuron

1-(3, 5-dichloro-2, 4-difluorophenyl)-3-(2, 6-difluorobenzoyl) urea

Teflubenzuron is an acylurea insecticide (Fig. **15**). Its major use is for the control of a wide range of insect pests from order Lepidoptera and Coleopteran and some mites in fruits, vegetables, cereals, nuts and seeds. These compounds are classified as benzoylphenylurea and possess a number of halogen substituent. Insects exposed to these compounds are unable to form normal cuticle because their ability to synthesize chitin (polysaccharide of N-acetylglucosamine) is absent.

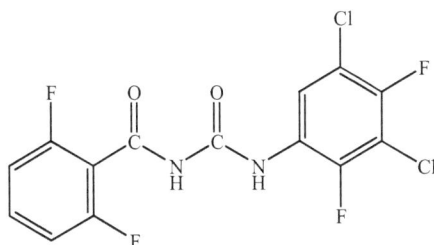

**Figure 15.** Structure of Teflubenzuron.

- **Trade or other names** : DART, DIARACT, NEMOLT, NOMOLT, CL 291,898, SAG 134

- **Appearance :** White to yellowish crystalline solid, odourless

- **Molecular formula** : $C_{14}H_6Cl_2F_4N_2O_2$

- **Molecular weight** : 381.1

- **Vapour pressure** : $8 \times 10^{-10}$ Pa at 20°C

- **Melting point :** 222.5°C

- **Solubility in water :** $2 \times 10^{-5}$ g/l at 20°C

- **Solubility in organic solvent:** Hexane, Methanol, Ethanol, Acetonitrile.

- **Breakdown in soil and groundwater**: In water, half-life at pH 5-7 is 30 days at 25°C. In aqueous solution teflubenzuron shows a half-life of approximately 10 days. In storage the half life is 2 years. Microbes are mainly responsible for degradation of teflubenzuron in soil. Under aerobic and anaerobic conditions, 3, 5-dichloro-2, 4-difluorophenylurea and 3, 5-dichloro-2, 4-difluoroaniline were the major products. Teflubenzuron shows practically no mobility once it is applied to the soil. This is attributed to their very low water solubility, very little leaching and high adsorption in soil.

### 2.5.1.4. Juvenile Hormone Mimics

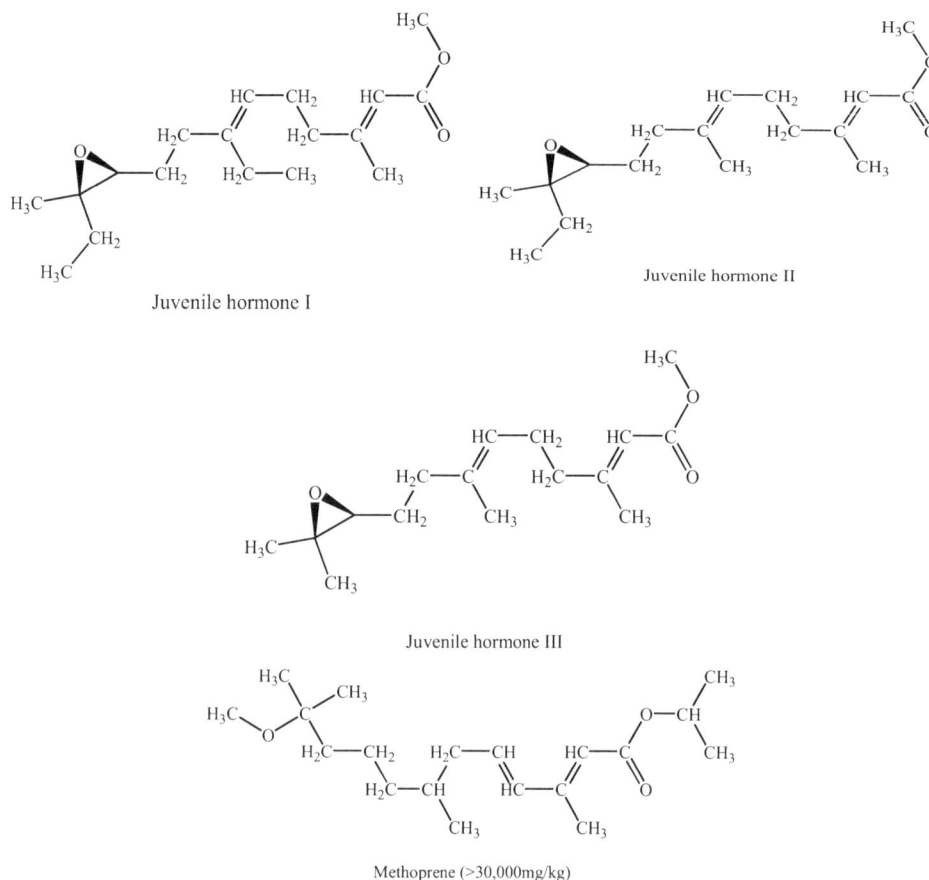

**Figure 16.** Structure of Juvenile Hormone Mimics.

The juvenile hormone mimics are compounds bearing a structural resemblance to the juvenile hormones of insects. They are lipophilic sesquiter penoids containing an epoxide and methyl ester groups (Fig. **16**). Two insecticidal mimics of juvenile hormone are *methoprene,* which bears a close structural resemblance to juvenile hormones, and *fenoxycarb,* which possesses a phenoxybenzyl group instead of a carbon chain with an epoxide. Both compounds are soluble in organic solvents and have extremely low toxicity to mammals. Exposure to these compounds at moulting results in the creation of insects containing mixed larval/pupal or larval/adult morphologies. Few important points associated with these compounds are,

- The efficacy of these compounds is highest when normal juvenile hormone titres are low, namely, in the last larval or early pupal stages.

- Timing of application is important for successful control.

- Another useful property of these compounds is that, in adults, they disrupt normal reproductive physiology and act as a method of birth control.

### 2.5.1.4.1. Fenoxycarb

Ethyl 2-(4-phenoxyphenoxy) ethylcarbamate

Fenoxycarb is non-neurotoxic carbamate insecticide and it blocks the ability of an insect to change into an adult from the juvenile stage (metamorphosis) (Fig. **17**). It also interferes with the moulting of larvae. It is an insect growth regulator used to control a wide variety of insect pests.

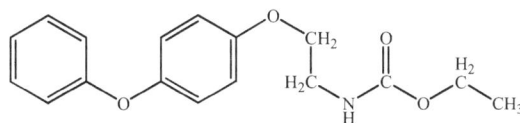

Fenoxycarb (>16,800 mg/kg)

**Figure 17.** Structure of Fenoxycarb.

- **Molecular Formula:** $C_{17}H_{19}NO$

- **Molecular Weight:** 301.30

- **Solubility in water:** 6mg/l

- **Toxicity, single dose:** The oral $LD_{50}$ of the compound is greater than 16,800 mg/kg in the rat. It is nearly non-toxic to mammals.

- **Breakdown in soil and water:** Fenoxycarb is readily broken down in soil by hydrolysis. Residues in soil were no longer detectable after three days of application.

### 2.5.1.4.2. Hydroprene

Ethyl (*E, E*)-(*RS*)-3, 7, 11-trimethyldodeca-2, 4-dienoate

Hydroprene is a growth regulator that inhibits the maturity and growth of certain insect pests in their immature stages (Fig. **18**). The chemicals interfere with the normal function of insect juvenile hormone, which controls the growth, development, and maturation of insects.

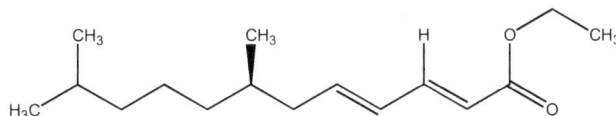

**Figure 18.** Structure of Hydroprene.

- **Trade or other name:** Gencor

- **Appearance:** Technical grade is an amber liquid.

- **Molecular Formula:** $C_{17}H_{30}O_2$

- **Molecular Weight:** 266.4

- **Solubility:** In water 2 mg/l. soluble in common organic solvents.

- **Toxicity, single dose:** $LD_{50}$ for bee is 1000μg / bee

- **Breakdown in soil and water:** Rapidly decompose in soil, half life is only few days.

## 2.6. BT TOXIN: BACILLUS THURINGIENSIS TOXIN

*Bacillus thuringiensis* (Bt) forms a crystalline inclusion body during sporulation that contains a number of insecticidal protein toxins. When consumed by the insect, the inclusion is dissolved in the midgut and releases δ-endotoxins. Mixtures of different δ-endotoxins are usually present in the inclusion and individual toxin proteins are designated with the prefix *cry*. The toxin proteins contain a few hundred to over 1000 amino acids. After they are ingested, the δ-endotoxins are cleaved to an active form by proteases within the midgut. The active toxins bind specifically to the membranes of the midgut epithelia and alter their ion permeability properties by forming a cation channel or pore. Ion movements through this pore disrupt potassium and pH gradients and lead to lysis of the epithelium, gut paralysis, and death. They are safe to vertebrates, invertebrates and other soil arthropods.

**Learning the Facts** : *Bacillus thuringiensis* (*Bt*), is an aerobic spore forming gram positive, rod-shaped soil bacterium was discovered in 1902 by S. Ishiwata, a Japanese microbiologist. First used as microbial insecticide against lepidopterous larvae in 1938. *Bt* toxins have contributed considerably to the field of insect pest management through novel, eco-friendly formulations and development of transgenic plants (fig. 19).

**Figure 19.** Bt protein crystals
Source : http://en.wikipedia.org/wiki/Bacillus_thuringiensis

**Structure of Bt Toxin** : The 3-domain toxins include proteins that are toxic to the insect orders Lepidoptera, Diptera, Coleoptera, and Hymenoptera. These are also known as Cry toxins. The primary action of Cry toxins is to lyse midgut epithelial cells in the target insect by forming lytic pores in the apical microvillar membranes. Cry toxin has three domains, each has its specific function. Domain I is located in the N-terminal part of the toxin molecule and makes pore in the cell membrane. Domain II in the form of hairpin structures with variable length. These structures are identified as specificity determining regions. Domain III is located in C-terminal part end of active toxin molecule. It involves in protecting the toxin against insect midgut proteases and also function as receptor binding, specificity determination and ion channel gating.

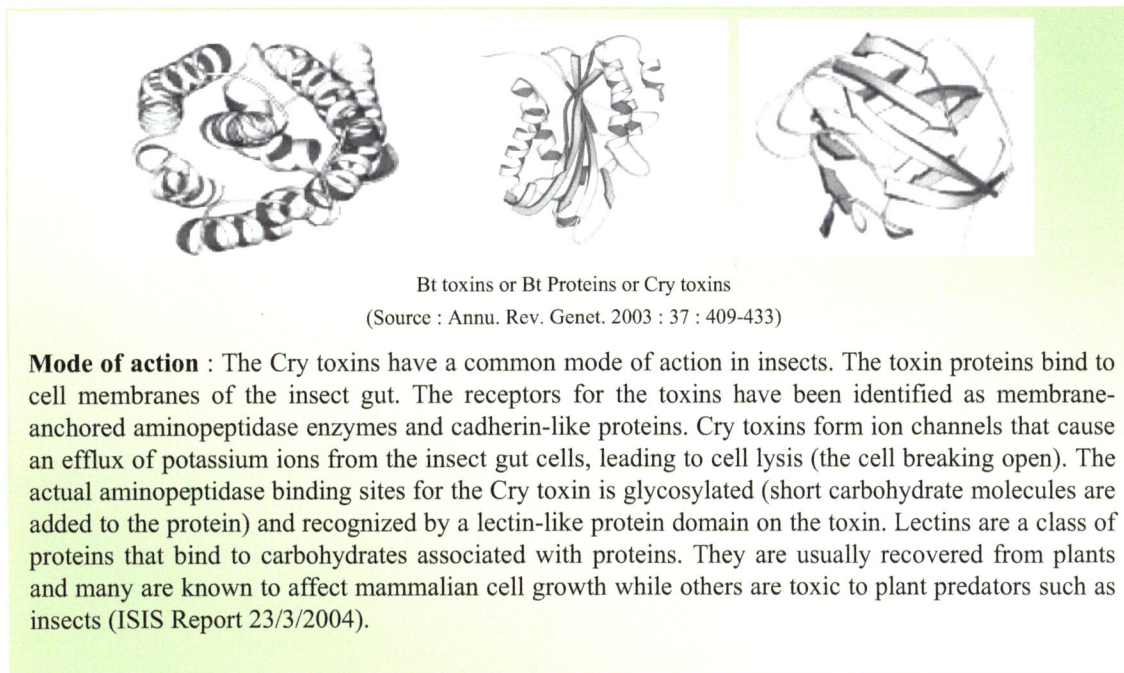

Bt toxins or Bt Proteins or Cry toxins
(Source : Annu. Rev. Genet. 2003 : 37 : 409-433)

**Mode of action** : The Cry toxins have a common mode of action in insects. The toxin proteins bind to cell membranes of the insect gut. The receptors for the toxins have been identified as membrane-anchored aminopeptidase enzymes and cadherin-like proteins. Cry toxins form ion channels that cause an efflux of potassium ions from the insect gut cells, leading to cell lysis (the cell breaking open). The actual aminopeptidase binding sites for the Cry toxin is glycosylated (short carbohydrate molecules are added to the protein) and recognized by a lectin-like protein domain on the toxin. Lectins are a class of proteins that bind to carbohydrates associated with proteins. They are usually recovered from plants and many are known to affect mammalian cell growth while others are toxic to plant predators such as insects (ISIS Report 23/3/2004).

## 2.7. ANTIFEEDANTS

Antifeedants are substances (aversive chemicals) which make the crop plant distasteful to the insect attempting to feed on it (Fig. **20**). Presence of antifeedant compound renders the host plant unpalatable to insect and inhibits the feeding. Some antifeedants inhibit feeding because they are toxic, others because they have a bad taste. They might also prevent insects to lay eggs. Bordeaux mixture is a feeding deterrent to fleas, beetles, leaf hoppers. Pymetrozin a pyridine azomethrine is an effective on aphids, leaf hoppers and whitefly. It is systemic and has long residual activity. It is non- injurious to natural enemies and the environment.

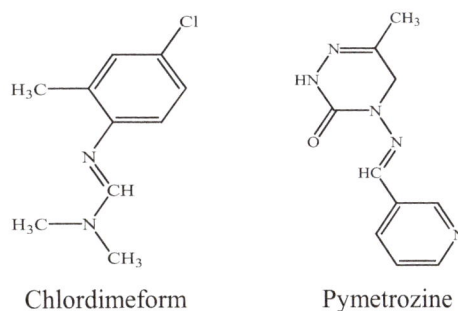

Chlordimeform                    Pymetrozine

**Figure 20.** Structure of Antifeedents.

Antifeedants can be found amongst all the major classes of secondary metabolites produced by the plants such as limonoids, quassinoids, diterpenes, sesquiterpenes, monoterpenes, coumarins, isoflavonoids, alkaloids, maytansinoids and ellagitannins. However, the most potent antifeedants belong to the terpenoid group, which has the highest number and diversity of known antifeedants.

### 2.7.1. Limonin

7, 16-Dioxo-7, 16-dideoxylimondiol

Limonin is chemically related to triterpene derivatives found in the *Rutacea*e and *Meliaceae* families named limonoids (Fig. **21**). All naturally occurring citrus limonoids contain a furan ring attached to the D-ring, at C-17, as well as oxygen-containing functional groups at C-3, C-4, C-7, C-16, and C-17. Limonin contains a C-14, 15-epoxide group (Somrutai *et al. EJEAFChe,* 4 (3): 938-944, 2005).

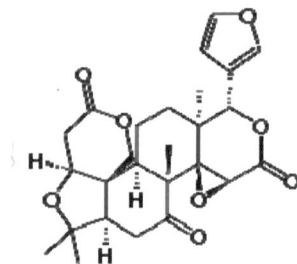

**Figure 21.** Structure of Limonin.

- **Other names:** Limonoate D-ring-lactone; Limonoic acid di-delta-lactone

- **Molecular Formula:** $C_{26}H_{30}O_8$

- **Molecular Weight:** 470.52

## 2.8. REPELLENTS

Repellents are chemicals which cause insects to orient themselves away from the source of chemical or its vapours thus preventing insect from reaching the target. Applying of these chemicals to skin, clothing, or other surfaces discourages insects (and arthropods in general) from landing or climbing on that surface. Common insect repellents include, oil of lemon, eucalyptus, picaridin or icaridin (a piperidine derivative), dimethyl phthalate, butopyronoxyl, citronella, DEET (Fig. **22**).

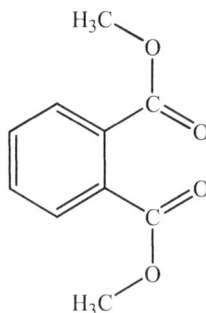

**Figure 22.** Structure of Dimethyl Phthalate.

Some insect repellents, particularly permethrin, are insecticides. Natural permethrin is known for its repellent properties. It is highly effective when applied to livestock to repel tsetse flies, and ticks. DEET is the most commonly used and most effective repellent against biting flies chiggers and certain species of mosquitoes. Insect repellents help to prevent and control the outbreak of insect-borne diseases such as malaria, Lyme disease, bubonic plague, and West Nile fever. Insects commonly serving as vectors for disease include fleas, flies, mosquitoes, and ticks.

## 2.9. ATTRACTANTS

Attractants are chemicals which causes the insects to orient themselves towards source or chemicals acting in vapour phase to move towards its source or zone of preferred concentration. Attractant can be of the following types *i.e.* poison baits, traps, oviposition lure.

- **Poison baits**: Attractants can be used in form of bait or in conjunction with a poison.

- **Traps**: It is a device for immediate killing of insects or retaining them for later destruction. Synthetic attractant such as medlure and trimedlure are used as attractant in traps for fruit flies.

- **Oviposition lure**: These are substances that attract gravid females and induce them to lay eggs. Houseflies are attracted to ammonia and green bottle flies are attracted to ammonium carbonate for oviposition.

## 2.10. SYNERGISTS

Synergists are compounds, when mixed with pesticides increases the toxicity to several folds. The combined effect of its exposure is considerably greater than the sum of the effects from the individual components. This phenomenon is known as synergism or potentiation. Synergism means, both chemicals have individual effects but the effect will be more when they come together. Potentiation means, one chemical has an effect and the second chemical has no effect but when they come together it enhances the effect of the former chemical on combined exposure for example sesame oil enhanced the insecticidal activity of pyrethrum and here the active compounds are sesamin and sesamolin. Effectiveness of insecticide synergist is expressed by LD50 of insecticide alone.

## 2.11. BOTANICAL PESTICIDES

Botanical pesticides are naturally occurring chemicals extracted from plants. Botanical pesticides are available as an alternative to synthetic chemical formulations but they are not necessarily less toxic to the humans. Some of them are lethal toxic, fast acting and potent carcinogens. Botanical pesticides break down readily in soil and are not stored in plant or animal tissue.

### 2.11.1. Citrus oil (limonene, linalool)

Are extracted from citrus peels and primarily used as flea dips, but have been combined with soaps as contact poisons against aphids and mites (Fig. **23**). They evaporate quickly after application and have no residues.

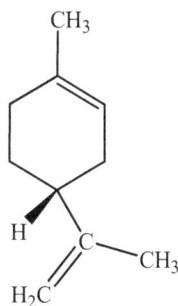

**Figure 23.** Structure of d-limonene.

### 2.11.2. Nicotine

Concentrate is very poisonous if inhaled. It is derived from tobacco and is commonly sold as a 40 percent nicotine sulfate concentrate (Fig. **24**). Nicotine is a fast acting contact killer for soft bodied insects, but does not kill most chewing insects.

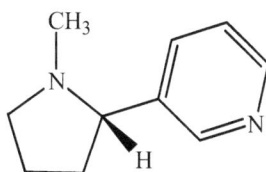

**Figure 24.** . Structure of Nicotine.

## 2.11.3. Pyrethrin

Is a fast acting contact poison derived from the pyrethrum daisy (Figs. **25** & **26**). It is very toxic to cold blooded animals. Pyrethrin is effective on most insects, but does not control mites. It rapidly breaks down in sunlight, air and water.

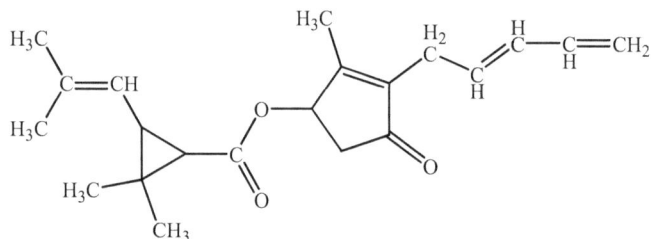

**Figure 25.** Structure of Pyrethrin I.

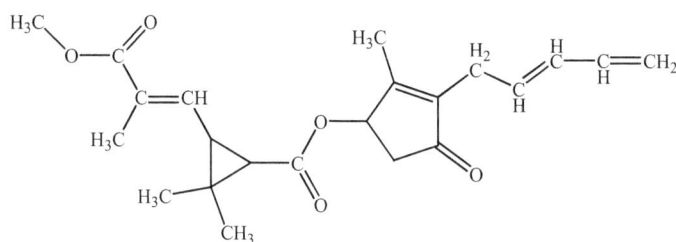

**Figure 26.** Structure of Pyrethrin II.

## 2.11.4. Rotenone

Is derived from the roots of over 68 plant species (Fig. **27**). It has a short residual time. Rotenone is a broad spectrum poison mainly used to control leaf-eating caterpillars and beetles.

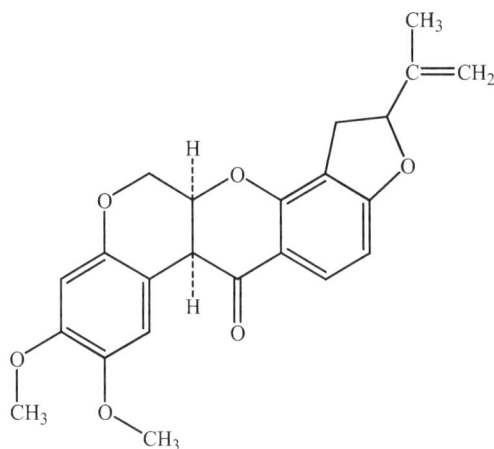

**Figure 27.** Structure of Rotenone.

## 2.11.5. Ryania

Is a slow acting stomach poison. It has a longer residual than most botanicals. Toxicity to mammals is moderate.

## 2.11.6. Sabadilla

Is derived from the seeds of South American lilies. It is a broad spectrum contact poison, but has some activity as a stomach poison. It is most effective against true bugs such as harlequin bugs and squash bugs.

## 2.11.7. Neem

Is a relatively new product on the market. It is derived from the neem tree that grows in arid tropical regions. Extracts from the neem tree have been reported to control over 200 types of insects, mites, and nematodes.

## 2.11.8. Azadirachtin

The insecticidal ingredient found in the neem tree is azadirachtin that belongs to an organic molecule class called tetranortriterpenoids (Fig. **28**). It is structurally similar to insect hormones called "ecdysones," which control the process of metamorphosis. Azadirachtin may act as "ecdysone blocker." It blocks the insect's ecdysones production and thus breaking their life cycle. It may also serve as a feeding deterrent for some insects. It is used to control whiteflies, aphids, thrips, fungus gnats, caterpillars, beetles, mushroom flies, mealybugs, leafminers, gypsy moths and others on food, greenhouse crops, ornamentals and turf.

1.  **Toxicity, single dose:** The acute oral toxicity in rats fed technical grade azadirachtin ranged from greater than 3,540 mg/kg to greater than 5,000 mg/kg,

2.  **Molecular Formula:** $C_{35}H_{44}O_{16}$

3.  **Molecular Weight:** 720

4.  **Chemical family:** Tetranortriterpenoids

5.  **Breakdown of Chemical in Soil and Groundwater:** Mobility in soil is very low. Accumulation in the environment is not expected.

6.  **Breakdown of Chemical in Surface Water:** A formulated product which contains the active ingredient azadirachtin is considered a water pollutant. It breaks down rapidly (in 100 hours) in water or light, and will not cause long-term effects.

7.  **Breakdown of Chemical in Vegetation:** Azadirachtin is considered non-phytotoxic when used as directed.

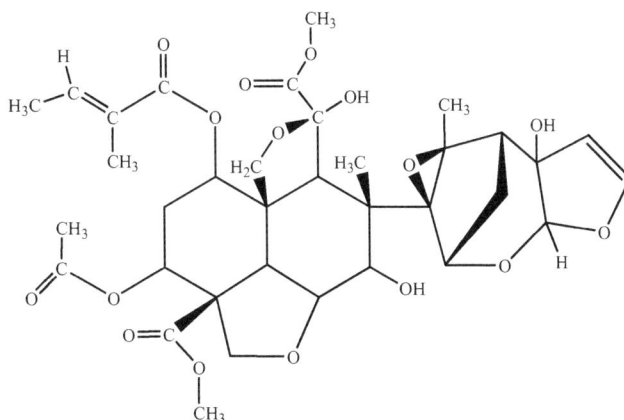

**Figure 28.** Structure of Azadirachtin.

## 2.12. BIOPESTICIDES

Biopesticides are derived from natural resources such as animals, plants, bacteria, and certain minerals. For example, canola oil and baking soda have pesticidal applications and are considered biopesticides. Biopesticides may be divided into three major classes:

## 2.12.1. Microbial pesticides

It consists of a microorganism (e.g., a bacterium, fungus, virus or protozoan) as the active ingredient.

- Microbial pesticides can control many different kinds of pests, although each separate active ingredient is relatively specific for its target pest[s]. For example, there are fungi that control certain weeds, and other fungi that kill specific insects, e.g. *Beauveria bassiana.*

- The most widely used microbial pesticides are subspecies and strains of *Bacillus thuringiensis, or Bt.* Each strain of this bacterium produces a different types of proteins, and specifically kills one or a few related species of insect larvae.

## 2.12.2. Plant Incorporated Protectants (PIPs)

Are toxic chemicals that are produced by the plant due to incorporation of genetic material from other source. Introduction of Bt gene in plant genome and its expression to produce Bt toxin that kills the specific larvae.

## 2.12.3. Biochemical Pesticides

Are naturally occurring substances that controls the pests by non-toxic mechanisms. It includes the substances, such as insect sex pheromones that interfere with mating as well as attract insect pests in traps.

## 2.13. NEONICOTINOIDS

The neonicotinoids are new class of insecticides developed in the past three decades (Fig. **29**). The mode of action of neonicotinoids is similar to the natural insecticide nicotine, which acts on the central nervous system. They are effective against sucking insects, but also chewing insects such as beetles and some Lepidoptera, particularly cutworms. Neonicotinoids commonly include acetamiprid, clothianidin, dinotefuran, imidacloprid, nitenpyram, thiacloprid, and thiamethoxam.

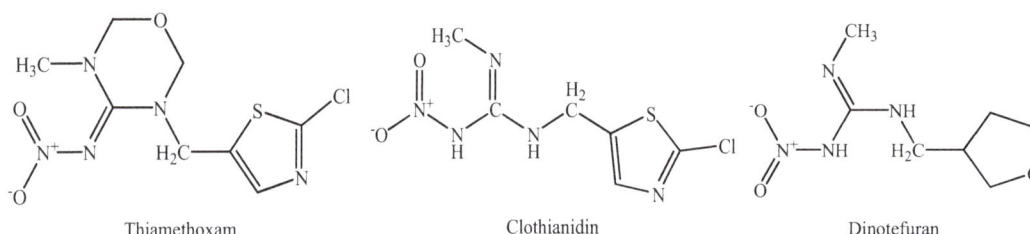

**Figure 29.** Structure of different neonicotinoids.

**Table 1.** Toxicity of different neonicotenoid pesticides ( $LD_{50}$ mg/kg body weight).

| Pesticide | Rat (oral application) | Rabbit (dermal application) |
|---|---|---|
| Acetamiprid | 450 | >2,000 (Tristar®) |
| Clothianidin | >5,000 | >2,000 (Acceleron®) |
| Imidacloprid | 4,870 (Gaucho®) | >2,000 (Admire®) |
| Thiamethoxam | >5,000 | >2,000 |

They have outstanding potency and systemic action for crop protection against piercing-sucking pests, and they are highly effective for flea control on cats and dogs. They are readily absorbed by plants and act quickly, at low doses, on piercing-sucking insect pests (aphids, leafhoppers, and whiteflies) of major crops. The neonicotinoids are poorly effective as contact insecticides and for control of lepidopterous larvae. They are used primarily as plant systemic, when applied to seeds, soil, or foliage they move to the growing tip and afford long-term protection from piercing-sucking insects, e.g., for 40 days in rice. IMI and nitenpyram are highly effective flea control agents on cats and dogs, and are administered as oral tablets or topical spot treatments while the nicotinoids are structurally similar to the neonicotinoids, they primarily differ by containing an ionisable basic amine or imine substituent.

### 2.13.1. Imidacloprid

($E$)-1-(6-chloro-3-pyridylmethyl)-$N$-nitroimidazolidin-2-ylideneamine

Imidacloprid is a neonicotinoid insecticide in the chloronicotinyl nitroguanidine chemical family (Fig. **30**). Neonicotinoid insecticides are synthetic derivatives of nicotine, an alkaloid compound found in the leaves of many plants in addition to tobacco (table 1). Imidacloprid is used to control sucking insects, some chewing insects including termites, soil insects, and fleas on pets. It may be applied to crops, soil, and as a seed treatment. It is a systemic insecticide that translocates rapidly through plant tissues following application. It acts on several types of post-synaptic nicotinic acetylcholine receptors in the nervous system.

- **Molecular Formula:** $C_9H_{10}ClN_5O_2$

- **Molecular Weight:** 255.7

- **Behaviour in Soil:** Soil half-life for imidacloprid ranged from 40 days in unamended soil to up to 124 days for soil recently amended with organic fertilizers. Metabolites found in agricultural soils are 6-hydroxynicotinic acid, (1-[(6-chloro-3-pyridinyl) methul]-2-imidazolidone), 6-chloronicotinic acid, with lesser amounts of a 2-imidazolidone.

- **Appearance:** Colorless crystals with a weak characteristic odor.

- **Water Solubility:** 0.51 g/l (200 ° C)

- **Solubility in Other Solvents:** at 20 ° C: dichloromethane - 50.0 - 100.0 g/l; isopropanol - 1.0-2.0 g/l; toluene - 0.5-1.0 g/l; n-hexane - <0.1 g/l.

- **Melting Point:** 136.4-143.8 ° C., 143.8 ° C (crystal form 1) 136.4 ° C (crystal form 2)

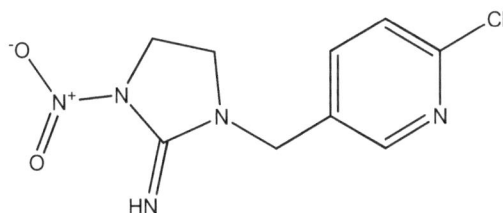

**Figure 30.** Structure of Imidacloprid.

### REFERENCES

[1]    The pesticide Mannual. Clive Tomlin, Ed., Crop Protection Publication, UK, 1997.
[2]    Agrochemicals. Franz Muller, Ed., WILEY-VCH, Federal Republic of Germany, 2000.
[3]    Pesticide Biochemistry and Physiology. C.F.Wilkinson (Ed.) Plenum Press, NY, 1976.
[4]    Dileep K. Singh, Toxicology of insecticides. http://nsdl.niscair.res.in/ CSIR, New Delhi, India, 2007.
[5]    Somrutai *et al.* Electron. J. Environ. Agric. Food Chem. 4 (3) : 938-944, 2005.
[6]    The Merck Index 13th Edition, 2001.
[7]    Biopesticides: Pest Management and Regulation. Alastair Bailey *et al.* (eds.) CABI International, 2010.

## QUESTIONS

1. Explain briefly, the organochlorine pesticide DDT. How it differs from Endosulfan?

2. Which one is better insecticide in reference to their persistence in the environment

   (i) DDT (ii) Carbaryl (iii) imidacloprid.

3. Write short notes on,

   (a) Bt toxin

   (b) Biopesticide

   (c) (c ) Neonicotinoids

   (d) Attractants

4. What is the difference between nicotenoids and neonicotinoids? Explain briefly their mode of action?

# CHAPTER 3

## Toxicology of Pesticides

**Abstract:** In this chapter, toxicity of molecules is described. The toxicity of pesticides to an organism is usually expressed in terms of the $LD_{50}$ (lethal dose 50 percent) and $LC_{50}$ (50 percent lethal concentration). And the interaction of toxic chemical with a given biological system is dose-related. So, it is the dose which makes substances poison. The right dose differentiates a poison and a remedy. At high doses, all the chemicals are toxic, at appropriate intermediates doses they are useful and at low enough doses they do not have a detectable toxic effect. There are certain pesticides (e.g. DDT) known today, because of the long term exposure to them at doses that do not immediately kill the organism showed severe effects like Carcinogenic, Mutagenic and Teratogenic effects.

Insects administered chemicals by several methods including *topical application, Injection Method, Dipping Method, Contact or Residual Method, Feeding and Drinking Method.* The susceptibility of insect population to a certain poison is assessed by constructing dosage-mortality curve in which the logarithmic scale of dosages is plotted against the probit units of percent mortalities at a given period of time.

**Keywords:** $LD_{50}$; $LC_{50}$; $ED_{50}$; NOAEL; LOAEL; Dose; Carcinogenicity; Mutagenicity; Teratogenicity; Mode of action.

## INTRODUCTION

Toxicology (from the Greek words *toxicos* means poisonous and *logos* means study) is the study of the adverse effects caused by chemicals on living organism. Here, pesticide is a toxic chemical. Toxicity of chemicals depends on the nature of toxicant, routes of exposure (oral, dermal and inhalation), dose and morphological and physiological state of organism. Toxicity of pesticides usually expressed in terms of $LD_{50}$ or $LC_{50}$. Its values are expressed in term of milligram per kilogram body weight or ppm respectively.

## 3.1. $LD_{50}$ (LETHAL DOSE FIFTY PERCENT)

The term $LD_{50}$ is expressed, as the single exposure dose of the substance per unit body weight of the organism, it is required to kill 50 % of the test the population under a defined set of conditions, where the population is genetically homogeneous. It is applied orally or topically under stated experimental conditions and usually expressed in terms of mg poison per kilogram body weight of the experimental animals (Table 1). Under certain conditions, the term micrograms per insect ($\mu$g/insect) may be used when chemical is applied topically to the insect. The $LD_{50}$ can be found for any route of entry or administration but dermal (applied to the skin) and oral (given by mouth) administration methods are the most common. It is a frequently used to measure of acute toxicity of an insecticide on an organism and is also important to know that the actual $LD_{50}$ value may be different for a given chemical depending on the route of exposure e.g., oral, dermal, inhalation. For example, some $LD_{50}$ in rat for dichlorvos, an insecticide commonly used in household pesticide strips, are listed below:

$LD_{50}$/ $LC_{50}$ (Rat)

1. Oral $LD_{50}$ :  56 mg /kg

2. Dermal $LD_{50}$ :  75 mg/kg

3. Intraperitoneal $LD_{50}$ :  15 mg/kg

In general, the smaller the $LD_{50}$ value, the more toxic is the chemical. The opposite is also true, larger the $LD_{50}$ value, the lower the toxicity (Table 1).

**Dileep K. Singh**
**All rights reserved - © 2012 Bentham Science Publishers**

**Table 1.** Toxicity Scale for Pesticides.

| Category | LD$_{50}$ oral mg/kg(ppm) | Examples |
|---|---|---|
| **Extremely toxic** | **1 mg/kg(ppm) or less** | **Parathion, aldicarb** |
| **Highly toxic** | **1-50 mg/kg(ppm)** | **Endrin** |
| **Moderately toxic** | **50-500 mg/kg(ppm)** | **DDT, Carbofuran** |
| **Slightly toxic** | **500-1000 mg/kg(ppm)** | **Malathion** |
| **Non-toxic (practically)** | **1-5 gm/kg** | |

**Learning the Facts :**

LD$_{50}$ is an abbreviation for Lethal Dose 50%, sometimes it is also referred as the 'Median Lethal Dose'. It was developed in 1927. Although the LD$_{50}$ is no longer the preferred method for estimating the acute toxicity of single doses of a substance, but for historical reasons it is still in use. In replacement to LD$_{50}$, the Fixed Dose Procedure (FDP) was proposed in 1984 by the British Toxicology Society. It is used to estimate acute oral toxicity of any substance. In this procedure, the substance is given to test organism (to five male and five female rats) at one of the four fixed-dose levels (5, 50, 500, and 2000 mg/kg body weight). Here, the objective is to identify a dose that causes a clear sign of toxicity but no mortality. Depending on the results obtained after first test, the experiment may be repeated for lower or higher doses. The results are thus interpreted in relation to animal survival and evident toxicity.

Differences in the LD$_{50}$ toxicity values are due to different routes of exposure. The LD$_{50}$ toxicity can be different for different animals. It gives a measure of the acute toxicity of a chemical or other substance in the strain, physiological status, sex, and age group of a particular test animal. Changing any of these variables may result into a different LD$_{50}$ value.

Once we have an LD$_{50}$ value, it can be compared to other LD$_{50}$ values by using a toxicity scale (Tables **2**, **3** & **4**). The two most common toxicity scales used are the "Hodge and Sterner Scale" and the "Gosselin, Smith and Hodge Scale". These tables differ in both the numerical values given to each class and the terms used to describe each toxicity class (Table **2**).

**Table 2.** Toxicity Classes: Hodge and Sterner Scale.

| | | Routes of Administration | | | |
|---|---|---|---|---|---|
| | | Oral LD$_{50}$ | Inhalation LC$_{50}$ | Dermal LD$_{50}$ | |
| **Toxicity ranking** | Terms | Single dose to rats (mg/kg body weight) | Exposure to rats for 4 hours (ppm) | Single application to skin of rabbits (mg/kg body weight) | Apparent Lethal Dose for a Man |
| **1** | Extremely Toxic | 1 or less | 10 or less | 5 or less | 1 grain (a taste, a drop) |
| **2** | Highly Toxic | 1-50 | 10-100 | 5-43 | 4 ml (1 tea spoon = 5 ml) |
| **3** | Moderately Toxic | 50-500 | 100-1000 | 44-340 | 30 ml (1 fl. oz. = 28.41 ml) |
| **4** | Slightly Toxic | 500-5000 | 1000-10,000 | 350-2810 | 600 ml (1 pint = 450 ml) |
| **5** | Practically Non-toxic | 5000-15,000 | 10,000-100,000 | 2820-22,590 | 1 litre (or 1 quart) |
| **6** | Relatively Harmless | 15,000 or more | 100,000 | 22,600 or more | 1 litre (or 1 quart) |

**Table 3.** Toxicity Classes: Gosselin, Smith and Hodge.

| Probable Oral Lethal Dose (Human) | | |
|---|---|---|
| **Toxicity Ranking** | Dose | For a Person ( 70 kg/ 150 lbs) |
| **1. Super Toxic** | Less than 5 mg/kg | 1 grain (a taste - less than 7 drops) |
| **2. Extremely Toxic** | 5-50 mg/kg | 4 ml (between 7 drops and 1 tsp = 5 ml) |
| **3. Very Toxic** | 50-500 mg/kg | 30 ml (between 1 tsp =5 ml and 1 fl ounce 30 ml) |
| **4. Moderately Toxic** | 0.5-5 g/kg | 30-600 ml (between 1 fl oz = 30 ml and 1 pint 450 ml) |
| **5. Slightly Toxic** | 5-15 g/kg | 600-1200 ml (between 1 pint = 450 ml to 1 quart =1 litre) |
| **6. Non-Toxic** | Above 15 g/kg | More than 1200 ml (more than 1 quart= 1 litre) |

Source: CCOHS OSH Answers, http://www.ccohs.ca/oshanswers/chemicals/ld50.html

This is another table used for pesticide toxicity (Table **3** & **4**).

**Table 4.** Toxicity Ranking Scale and Labeling Requirements for Pesticides.

| Group | Label | $LD_{50}$ oral mg/kg(ppm) | $LD_{50}$ dermal mg/kg(ppm) | Apparent oral lethal dose |
|---|---|---|---|---|
| **I** **Highly** **toxic** | DANGER-POISON marked with (skull and crossbones) | less than 50 | less than 200 | a few drops to a teaspoon (5 ml) |
| **II** **Moderately** **toxic** | WARNING | 51 to 500 | 200 to 2,000 | over 1 teaspoon 5 ml to 30 gm (1 ounce) |
| **III** **Less toxic** | CAUTION | over 500 | over 2,000 | over 30 gm (1 ounce) |
| **IV** **Non-toxic** | not required | --- | ---- | --- |

In Europe, $LD_{50}$ no longer in use to classify new chemicals for toxicity. Tests use the survival rate instead of mortality. Toxicity based on this approach will provide the same toxicity classification as the old one. Here suffering to test animal is less in comparison to traditional $LD_{50}$ value.

### 3.2. LC$_{50}$ (LETHAL CONCENTRATION FIFTY PERCENT)

$LC_{50}$ is the concentration of the chemical in the external medium (usually air or water surrounding experimental animals), which causes 50 % mortality of the test population, where the population is genetically homogeneous. This value is used when the exact dose given to the individual is difficult to be determined. $LC_{50}$ is expressed as the percent of active ingredient of the chemical in the medium or as parts per million (ppm). Example: Dichlorvos (for rat).

1.   Inhalation $LC_{50}$:   1.7 ppm (15 mg/m$^3$), 4-hour exposure

### 3.3. ED $_{50}$ (MEDIAN EFFECTIVE DOSE)

Statistically derived dose of a chemical expected to produce a certain effect in 50% of test organisms in a given population or to produce a half-maximal effect in a biological system under a defined set of conditions.

### 3.4. NOAEL AND LOAEL

NOAEL (no observed adverse effect level) and LOAEL (lowest observed adverse effect level) are used for setting regulatory levels such as reference doses (RfDs), reference concentrations (RfCs), and acceptable daily intakes (ADIs). The common definition of NOAEL is "the highest experimental point that is without adverse effect". It can be further explained as the highest level of continual exposure to a chemical which causes no significant adverse effect on morphology, biochemistry, functional capacity, growth, development or life span of individuals of the target species used in the toxicology studies. The lowest-observed-adverse-effect level (LOAEL) is the lowest concentration or amount of a chemical, which causes an adverse effect on morphology, function, growth, development and life span of a target organism under defined conditions of exposure (Fig. **1**).

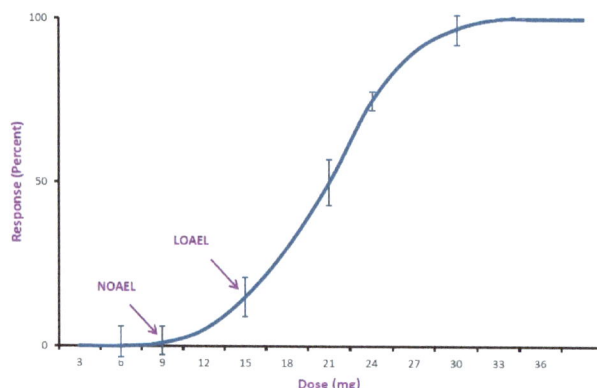

**Figure 1.** NOAEL: It is the highest data point at which no observed adverse effect was found. LOAEL: It is the lowest data point at which observed adverse effect was found
Source: http://www.eoearth.org/article/Dose-response_relationship

### 3.5. DOSE RESPONSE RELATIONSHIPS

Toxicity of chemical is determined by quantifying the response on test animal to a series of increasing dose. The relationship between animal and administered dose can be graphically presented as dose-response relationship curve. Dose response relationships means the relationship between the dose of a chemical substances administered or received and the incidence of an adverse health effect in exposed population (Fig. **2**).

In toxicology, the dose is very important which determine its impact on organism. It is the dose which makes substances a poison. The right dose differentiates a poison and a remedy. At high doses, all the chemicals are toxic, at judicious doses they are useful and at very low doses they do not have a detectable toxic effect. A dose-response relationship is based on the following important assumptions:

1.  There is always a threshold dose below which no effect occurs.

2.  Once effect occurs, response increases as dose increases.

3.  Once a maximum response is reached, any further increases in the dose will not result in any increased effect.

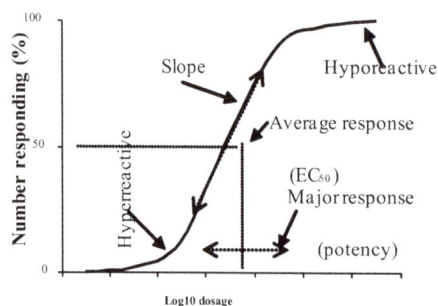

**Figure 2.** Typical sigmoid cumulative dosage-response curve for a toxic effect which is Symmetrical about the average (50 percent response) point.

When a genetically homogeneous population of animals of the same species and strain is exposed to a toxicant, the proportion exhibiting a particular toxic effect will increase as the dosage increases. This is shown schematically (Fig. **2**) as a cumulative distribution curve, where the number of animals responding is plotted as a function of the dosage given (as a $\log_{10}$ function). A few individuals respond to relatively low doses (constituting a hyperreactive group), most respond to medium dose, and a small number required a relatively high dose (constituting a hyporeactive group) before they are affected. The lowest dose at which an effect is discernable is referred to as the no observable effect level (NOEL).

## 3.6. CARCINOGINICITY, MUTAGENICITY AND TERATOGENICITY

To understand the effects of pesticides on organisms, a study on long term exposure of pesticides at the doses that do not immediately kill the organism are required. These are referred as long-term, or chronic, studies. The consequences of chronic exposure to toxicants can results in different responses in the exposed organisms. The following responses are typically evaluated for pesticides.

### 3.6.1. Carcinoginicity

Carcinogenesis is the production or increase in cancer frequency in the test organisms relative to exposure to a toxicant.

1. A number of pesticides have been reported to induce tumours in mice and rats in laboratory tests.

2. DDT and other chlorinated hydrocarbon insecticides have been shown to cause marked changes in the liver of various rodents, and these changes may progressed to tumour formation in some species, notably in the mouse. However, DDT failed to produce detectable tumours in pesticide industry workers who absorbed DDT for 19 years or more at rates hundreds of time higher than those found in the general population.

### 3.6.2. Mutagenicity

Mutagenicity refers to the induction of permanent changes in the amount or structure of genetic material of cells or organisms, which can be transmitted to the next generations. Changes induced in cells by a mutagen can cause cancer, while damage to the egg and sperm can cause adverse reproductive and developmental outcomes.

> **Learning the facts**: The use of the Ames test is based on the assumption that any substance that is mutagenic may also turn out to be a carcinogen. Although, some substances that cause cancer in laboratory animals ( e.g. dioxin) do not give a positive Ames test. The bacterium used in the test is a strain of *Salmonella typhimurium* that caries a defective (mutant) gene making it unable to synthesize the amino acid histidine (His) from the ingredients in its culture medium. However, some types of mutations can be reversed, a back mutation, with the gene regaining its function.

1. There are a large number of assays for mutagenesis such as point mutations in bacteria, yeast and mold, mammalian cells *in vitro*, tests for chromosomal aberrations, *etc.*

2. The Ames Test with *Salmonella typhimurium* is a popular assay procedure and is used extensively by many investigators and chemical industries to test the safety of new products before their introduction in the market.

### 3.6.3. Teratogenesis

Teratogenesis is a medical term from the Greek, literally meaning *monster-making*, which derives from teratology, the study of the frequency, causation, and development of congenital malformations misleadingly called *birth defects*.

1.  Teratogenesis has gained a more specific usage for the development of abnormal cell masses during foetal growth causing physical defects in the foetus. Such deformities can also be observed in rodent foetuses by direct inspection or with the aid of a microscope.

2.  Teratogenesis does not include toxic injury to organs after they are fully formed.

## 3.7. METHODS FOR ESTIMATING TOXICITY OF CHEMICALS

There are several methods of administering a chemical to an insect. Commonly employed methods are,

### 3.7.1. Topical Application

Where the insecticide is dissolved in a relatively nontoxic and volatile solvent such as acetone, and is then allowed to come in contact with a particular location on the body surface. The results are expressed as micrograms of active ingredient per insect ($\mu$g AI/ insect) or ($\mu$g AI/g insect). The advantages of this method are,

1.  The high degree of precision and reproducibility that can be attained.

2.  The large number of tests that can be performed in a relatively short time.

3.  The small number of insects (10-20) required per replication.

4.  The simple and inexpensive equipment needed.

5.  The very small amount of chemicals and solvents used.

6.  The fact that the $LD_{50}$ values obtained for any species are reasonably constant and reproducible from laboratory to laboratory, provided that identical conditions of testing are maintained.

### 3.7.2. Injection

Injection method is used when knowledge of the exact amount of insecticide inside the body of the insect is required.

1.  Very fine stainless steel needles of 27 or 30 gauges (0.41 or 0.30 mm in diameter) are used. Small glass needles of 0.1-0.16 mm in diameter may be used for injection to small insects.

2.  The insecticide is commonly dissolved in propylene glycol or peanut oil and injection is made intraperitoneally (into the body cavity).

3.  Care must be taken to avoid bleeding by the insects.

### 3.7.3. Dipping

This method is employed when topical application or injections are impractical, e.g., with small plant feeding insects, housefly larvae, insect eggs, *etc.*

1.  The insects are dipped in aqueous solutions, emulsions, or suspensions of the chemical for short periods of time.

2.  In this case, the $LC_{50}$ is used to express the results.

### 3.7.4. Contact or Residual

The insecticide in a volatile solvent is applied to a glass container such as a vial or a jar. The solvent is allowed to evaporate by rotating the container so that the insecticide is spread evenly over the entire surface leaving a residual film. Alternatively, the insecticide is applied evenly on a glass, filter paper, wood panel or other types of building materials and allowed to dry before exposing the insects to the residual deposits. The deposits are expressed as mg or g of active ingredient per square meter (mg or g AI/m$^2$).

### 3.7.5. Feeding and Drinking

These methods are used to evaluate the toxicity of ingested chemicals. They may be classified as unlimited availability of food or drink or as limited dose feeding.

1.   Unlimited feeding includes: textiles tests for moth-proofing, treated flour or grain, media for fly larvae, sprayed or dusted foliage, poison baits.

2.   Limited dose feeding include: coated leaves disks, sandwiches, squares, strips and pellets.

3.   Drinking methods include: sugar syrups and drinking through membranes such as plant juices or blood. In addition, there are methods for screening chemical attractants and repellents, for screening animal sprays, dusts, dips and dressings, and techniques for evaluating systemic insecticides against livestock insects.

### 3.8. ESTIMATING TOXICITY ON INSECTS AND OTHER ANIMALS

The toxicity evaluation process or toxic interactions of any chemical and any given biological system are dose related. At extremely high concentrations, most chemicals have toxic effects on biological systems. The toxicology of poisonous chemicals can be termed as the science of doses.

In order to assess the susceptibility of any population to a certain poison, probit units of percent mortalities are customarily plotted against a logarithmic scale of dosages. This method of computation yields a straight line which facilitates the determination of the LD$_{50}$ and other values on the plot (Fig. 3).

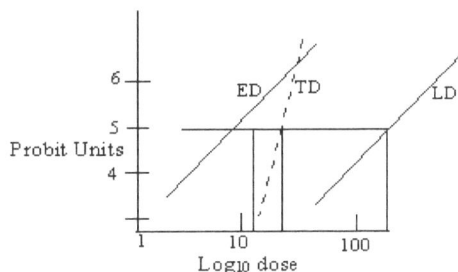

**Figure 3.** Plot of log $_{10}$ doses versus probit value of percent mortality showing effective dose (ED), toxic dose (TD), and lethal dose (LD).

If there is natural mortality in the controls, adjusted mortality is used according to Abbott's formula (Abbott 1925) as follows:

$$\text{Corrected Mortality Percent} = \frac{(P - P_0)}{(100 - P_0)} \times 100$$

Where P is the percent mortality of treated insects and P$_0$ is the percent mortality of insects in the untreated control. This adjusted value is permissible when mortality in the controls does not exceed 20 percent or when mortality is based on a large number of replications.

Evaluation of toxicity in higher animals is different from that of insects because the number of available animals for testing usually is limited. While the process for determining LD$_{50}$ is identical, greater emphasis is placed on qualitative rather than quantitative aspects of poisoning. Another characteristic of toxicological tests in higher animals is that, in most cases, the overriding concern is the evaluation of safety for man.

Selection of the test animal is usually based on convenience and cost. For ordinary testing of $LD_{50}$ values rats or mice are the animals of choice. The animals should be healthy and of acceptable genetic homogeneity. Factors influencing toxicity includes, duration of exposure, route of administration, species, individual variation, age, sex, population density, temperature and nutrition.

## 3.9. MODE OF ACTION OF PESTICIDES

Majority of the pesticides attack the nervous system. This shows irreversible damage, more so than any other tissue in the body. Other chemicals, whose primary target is elsewhere, may also produce their ultimate effect on the nervous system. For example, the heart poisons like atropine and poison which block the oxygen-carrying capacity of blood like carbon monooxide are lethal because they damage the brain damage by depriving it from its immense oxygen requirement. Understanding the mechanism of action of pesticides is a major and fundamental task for pesticide toxicologists and pharmacologists.

In order to understand how insecticides exert their toxic effects, it is essential to have some fundamental understanding of the physiology and biochemistry of the mammalian and insect nervous system.

### 3.9.1. Conduction of Nerve Impulse

The nervous system is composed of billions of specialized cells called neurons. The cell body or soma of a neuron contains mitochondria, ribosomes, a nucleus, and other organelles (Fig. **4**). It has dendritic branches which serve as receptor sites for information sent from other neurons. If the dendrites receive a strong enough signal from a neighbouring nerve cell, or from several neighbouring nerve cells, the resting electrical potential of the receptor cell's membrane becomes depolarized. Regenerating itself, this electrical signal travels down the cell's axon, a specialized extension from the cell body which ranges from a few hundred micrometers in some nerve cells, to over a meter in length in others. This wave of depolarization along the axon is called an action potential. Most axons are covered by myelin, a fatty substance that serves as an insulator and thus greatly enhances the speed of an action potential. In between each sheath of myelin is an exposed portion of the axon called a node of Ranvier. It is in these uninsulated areas that the actual flow of ions along the axon takes place.

There are two different modes of transmission in the nervous system,

1. Axonal transmission

2. Synaptic transmission.

### 3.9.1.1. Axonal Transmission

Axon is the part of the neuron, which is specialised for carrying nerve impulses or action potentials rapidly without changing the size or pattern of the impulse as it moves along.

1. Under normal conditions, extracellular fluid surrounding the axonal membrane has a high concentration of $Na^+$ and a low concentration of $K^+$ ion. Hence neurons possess a transmembrane voltage of about -60mV on the inner side of the cell. This is known as the *Resting Membrane Potential (RMP)*.

2. A stimulation causes axonal membrane to become permeable to $Na^+$, this causes sodium influx resulting inside membrane transiently positive. This constitutes the rising phase of the action potential. However, the sodium channels start closing quickly, usually within 1m sec. The membrane now becomes permeable to $K^+$ ion, and because of its higher concentration in inside, rushes out (potassium efflux), constituting the falling phase of the action potential.

3. The ability to perform all these events depends upon the maintenance of gradient across the membrane. This is done by $Na^+/ K^+$ pump which maintain the gradient in the first place and secondly compensates for the leakage that occurs during impulse transmission.

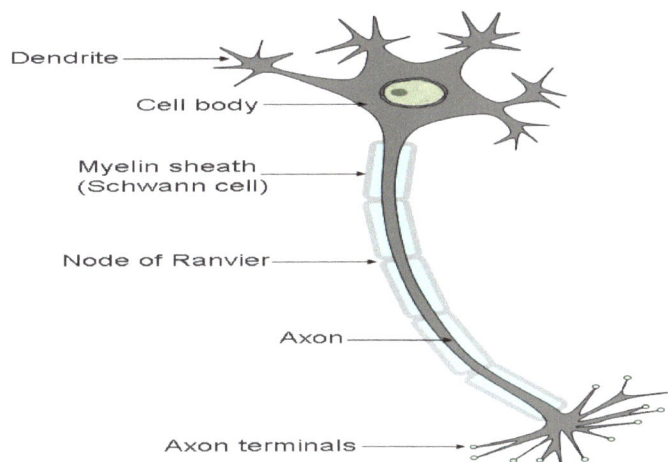

**Figure 4.** Structure of nerve cell
Source: http://scienceblogs.com/neurotopia/2006/07/stem_cells_for_spinal_cord_inj.php

### 3.9.1.2. Synaptic Transmission

When an impulse has passed along an axon, it must cross a synapse to stimulate another neuron. Transmission across a synapse involves a chemical transmitter which is stored in vesicles in the end of the axon (unlike transmission across an axon, which is electrical transmission). The transmitter becomes attached at receptor sites to the post-synaptic membrane. This causes a change in the ion permeability, which leads to membrane depolarisation generating an action potential.

The two most common types of neurotransmitters are,

1. Acetylcholine and

2. Norepinephrine

    (a) Universally, the synapse which utilizes acetylcholine is called *cholinergic,* while the one utilising norepinephrine is called *adrenergic synapse.*

    (b) In order to restore the sensitivity of the synapse, the transmitter must be eliminated, so that the receptor can return to its resting state.

    (c) At cholinergic junctions this is done by cholinesterase which hydrolyses acetylcholine into inactive components, choline and acetate.

    (d) At adrenergic junction, the corresponding degrading enzyme is monoamino-oxidase.

### 3.9.2. Central Nervous System (CNS)

It refers in mammals to the brain and the spinal cord and in insects to the chain of ventral ganglia. The brain is the integrating centre of all body activities. The peripheral nervous system consists of the somatic and the autonomic systems.

1. The somatic system handles those movements characterised by reaction to environmental stimuli and the corresponding muscle response. Transmission across the synapse and at the neuromuscular junction is cholinergic.

2. The autonomic division of the nervous system innervates all the effectors of the body except the skeletal muscles. There are two divisions in the autonomic nervous system, the

sympathetic and the parasympathetic. Sympathetic system has adrenergic system while parasympathetic system uses cholinergic system. Beside these there are many other chemicals which act as transmitters in the central nervous system. Gamma-aminobutryic acid (GABA) is one of them.

Insect nervous system is analogous to that of mammals. There are however some differences. Firstly, there are histochemical, enzymological and physiological evidences those neuromuscular junctions of insects are not cholinergic. Instead Glutamic acid stimulates and GABA suppresses muscle contraction as transmitters at the neuromuscular junction. Also there is no distinct autonomic system. Insect nerve also shows no distinct myelination.

Acetylcholine was established as the transmitter at the insect nervous system synapse, although GABA, glutamic acid, glycine, biogenic amines such as dopamine, norepinephrine, serotonin and tryptaminem do occur in insect nervous system. Octopamine is unique to the insect system.

### 3.9.3. GABA Receptor Complex

GABA receptor complex or GABA receptor chloride ionophore is a single complex protein having atleast three distinct interacting components both in mammals and in insects. GABA is released from the presynaptic endings of stimulated inhibitory neurons and diffuses across the synaptic cleft. Now it binds to the GABA receptor complex at the postsynaptic site. This increases the chloride permeability of the axon thus causing hyperpolarisation of the nerve fibre. All these events lead to the deactivation of the postsynaptic cell.

Thus activation of GABA receptors produces inhibition at a variety of sites, while inhibition of GABA-induced chloride ion permeability causes excessive release of acetylcholine at presynaptic sites or stimulates glutamine action, which may account for stimulant and convulsion effects of the inhibitors.

### 3.9.4. Mode of Action of Organochlorine Insecticides

### 3.9.4.1. DDT

DDT is a slow acting neurotoxicant. It shows a negative correlation with temperature i.e. its insecticidal potency increases with the decrease in temperature. Its exact biochemical mechanism has not been elucidated yet although it is suggested that the ATP-dependent portion of the NA+/Ca+ exchange may be involved. It is now established that DDT acts primarily on neurons and interferes with the axonal transmission. DDT prolongs the closure of sodium gated channels, thereby increasing the depolarisation after-potential. When this has increased to a certain level, a sudden burst of repetetive discharge or a trail of impulse is provoked by a single stimulus. This leads to hyperexcitability of the nervous system resulting in tremors, paralysis and even death. DDT has been shown to cause the release of neurohormones which might be involved in its toxicity.

### 3.9.4.2. Hexachlorocyclohexane (HCH)

Among its isomers, only gamma HCH or lindane has high toxicity towards insects and other organisms. It is a more acute nerve poison than DDT. There is negative correlation of its toxicity with temperature, but not as pronounced as in DDT poisoning. Practically nothing is known about the biochemical basis of insecticidal action of lindane. It is a better inhibitor of $Na^+$, $K^+$ and $Mg^+$ ATPase than DDT. Lindane causes accumulation of acetylcholine in nerves of insects but it does not inhibit the enzyme cholinesterase. The mechanism of action was established as blocking of the GABA-gated chloride channels.

### 3.9.4.3. Cyclodiene Insecticides

Like most organochlorine inscticides, cyclodienes also are neurotoxicants. But unlike other organochlorine insecticides they show a positive correlation with temperature *i.e.* their toxicity is enhanced with increase in temperature. There is a characteristic 'lag period' between the administrations of the poison and the appearance of poisoning symptoms in case of cyclodiene insecticides. Cyclodienes cause an excessive release of acetylcholne, but do not block the enzyme cholinesterase (AChE). There is evidence that they interact with ATPases from nerve cord and muscle. Cyclodiene compounds are particularly more dangerous because of their high oral and dermal toxicity.

### 3.9.5. Mode of Action of Organophosphorous Insecticides

Organophosphorous [OP] insecticides have structural complementarity with AChE enzyme, thus they mimic the gross molecular shape of acetylcholine. OPs react with a serine hydroxyl group within the enzyme active site, phosphorylating this hydroxyl group and yielding a hydroxylated leaving group (figure). This process inactivates the enzyme and blocks the degradation of the neurotransmitter acetylcholine. The synaptic concentrations of acetylcholine then build up and hyperexcitation of the CNS occurs. The signs of intoxication include restless, hyperexcitability, tremors, convulsions, and paralysis. In insects, the effects of OPs are confined to the CNS, where virtually all of the cholinergic synapses are vacuolated. The phosphorylation of acetylcholinesterase by OPs is persistent, reactivation of the enzyme can take many hours or even days.

### 3.9.6. Mode of Action of Carbamate Insecticides

Carbamates react with acetylcholinesterase in the same manner as that of OP compounds. They also bind to the enzyme cholinesterase forming a reversible complex. In this case, the reaction yields a carbamylation of the serine hydroxyl group. Complex decomposes into stable carbamylated enzyme (enzyme rendered inhibited) and a hydroxylated leaving group. Finally carbamylated enzyme is hydrolysed to regenerate the free enzyme and methylcarbanic acid. Only difference between the two group of insecticides is that phosphorylated enzyme (in the case of OP compounds), hydrolyses at a much slower rate as compared to the carbamylated enzyme (in the case of carbamates). Thus animals showing carbamate poisoning recover within hours after exposure to carbamates, unlike the OP compounds.

### 3.9.7. Synthetic Pyrethroids

Pyrethroids are typically esters chrysanthemic acid having a high degree of lipophilicity (fat solubility). Pyrethroid mode of action is classified as Type 1 or Type 2, depending on the poisoning symptoms.

#### 3.9.7.1. Pyrethroids

This group includes a non-alpha-cynopyrethroids, including natural pyrethrins, allethrin, tetramethrin, etc. They are characterised by whole body tremors similar to that in DDT. These insecticides prolong the sodium current during excitation, cause depolarisation after potential to increase. When the after potential exceeds the membrane threshold, repetitive action potentials are generated, leading to hyperexcitation. This is followed by tremors, paralysis and even death (Type 1). Another group (Type 2) includes alpha-cynopyrethroids including cypermethrin, deltamethrin, fenvalerate. They produce a distinctly different syndrome, characterised by sinus writhing, convulsions, accompanied by profuse salivation. They also act on the sodium channels, prolonging the sodium current to a greater extent than type1. Thus they depolarize the nerve membrane more strongly than type1, because of membrane depolarisation, nerve fibres do not initiate repetitive discharges, but sensory neurons discharge bursts of impulses and synaptic transmission is disturbed. The nerve conduction is eventually blocked due to membrane depolarisation. At higher concentrations type 2 pyrethroids bind to the chloride ionophore component of the GABA receptor complex and inhibit the GABA dependent chloride flux.

### 3.9.8.. Representative Mode of Action

#### 3.9.8.1. Metabolic Inhibitors

Certain chemicals affect the electron transport chain thus disrupting the normal metabolic pathway. Examples are rotenone (slows heartbeat, depresses respiration and oxygen consumption, and causes paralysis and death) and arsenicals (inhibit respiratory enzymes).

#### 3.9.8.2. Muscle Poisons

Certain chemicals have a direct action on the muscle tissue. Examples are Ryania and Sabadilla which increases oxygen consumption, followed by paralysis and death.

### 3.9.8.3. Alkylating Agents

Certain chemicals react directly with chromosomes and enzymes in the cells. Examples are fumigants such as methyl bromide and ethylene dibromide.

### 3.9.8.4. Physical Toxicants

Certain chemicals mechanically block the physiological processes. Examples are oil (blocks respiratory openings in insects) and boric acid and silica gel (effects insect cuticle causing dehydration and death).

### 3.9.8.5. Cytolytic (Cellular) Toxins

Certain chemicals cause cells to rupture and disintegrate. Example is Bt toxin from *Bacillus thuringiensis,* which is ingested by insect larvae and disrupts cells in the gut (causing paralysis of gut and cessation of feeding).

## REFERENCES

[1]    Perry A.S., Yamamoto I., Ishaaya I. and Perry R. Insecticides in agriculture and environment. Narosa Publishing House, New Delhi, India 1998.

[2]    Anderson D. and Conning D.M. Experimental Toxicology: The Basic Issue. (2nd Ed.) Royal Society of Chemistry, UK, 1993.

[3]    Helmut Greim and Robert Synder. Toxicology and risk assessmenT: A comprehensive introduction. John Wiely & Sons, Ltd. 2008

[4]    Robert C. Smart and Earnest Hodgson. Molecular and biochemical toxicology. John Wiely & Sons Ltd. 2008

## QUESTIONS

1.    Define the following terms,

   (a)   $LD_{50}$

   (b)   $LC_{50}$

   (c)   $ED_{50}$

2.    Explain briefly the mode of action of DDT and how it differs with organophosphorus pesticides?

3.    Write a short notes on,

   (a)   Cholinesterase inhibitors

   (b)   GABA receptor

   (c)   Corrected mortality

# CHAPTER 4

## Metabolism or Degradation of Pesticides: Phase I and Phase II Reactions

**Abstract:** In this chapter pesticide metabolism has been described. Pesticide Metabolism is typically a two stage process. These are Phase I and Phase II reactions. Phase I reactions normally add a functional (polar reactive) group to the foreign molecule which enables the phase 2 reaction to take place. These reactions are catalyzed by the cytochrome P450 group of enzymes and other enzymes which are associated with endoplasmic reticulum. Phase I reactions includes, Microsomal oxidation and Extramicrosomal oxidation reactions. Phase II reactions are conjugation reactions and involve the covalent linkage of the toxin or phase I product to a polar compound. In general, conjugated products are ionic, polar, less lipid soluble, less toxic and easily excretable from body.

All reactions of pesticide metabolism are described using chemical structure with possible target sites and enzyme involved.

**Keywords:** Phase I reaction; Phase II reaction; Microsomal oxidation reactions; Cytochrome P450; NADPH Cytochrome c reductase; Epoxidation and aromatic hydroxylation; O-dealkylation; N-dealkylation; S-dealkylation; Desulfuration; S-oxidation; Reduction reactions, Hydrolysis; Conjugation reactions; Methylation.

## INTRODUCTION

The toxicity and persistence of any pesticide in the body of organism depends on the ability of the organism to metabolized and excrete the pesticide. This is a self defence mechanism used by the organisms for their survival. Metabolism is in general considered as the detoxification process. However, there are examples where metabolism leads to more toxic metabolite. Such process is known as bioactivation of the pesticide. Metabolism and metabolites helps us in understanding the resistance mechanism developed by insects to various pesticides. This is useful in understanding the development of resistance in insects for the pesticides and in resistant management.

Most xenobiotics compounds (pesticides) are lipopholic in nature. This helps them to bind and penetrate the lipid cell membrane and transported by lipoproteins in the blood. Also being non soluble in water, they are not easily excreted from the body, until and unless they are converted to some polar compound. Therefore, the first step in any xenobiotic metabolism is to convert the compound into a more or less polar one so that it can be easily excreted from the body.

Williams (1959) first suggested that the metabolism of xenobiotic compound usually occur in two phases,

### 1. Phase I Reactions

In phase I reactions, the lipophilic xenobiotic compounds are converted into polar compound by introducing a polar group into it. In phase I reactions, a compound undergoes Oxidation, Reduction or Hydrolysis. The product of phase I reaction may serve as a substrate for Phase II reactions.

### 2. Phase II Reactions

In phase II reactions, the xenobiotic compound or the phase I product undergoes conjugation reaction with various endogenous molecules like sugars, amino acids, glutathione, phosphate and sulphate. Conjugation products are usually more polar less toxic and more easily excreted.

### 4.1. PLACE OF METABOLISM

Metabolism of xenobiotic compounds (pesticides) occurs in all organs and tissues of the organism. Here, the metabolism is considered as a defence mechanism. The purpose of it is to speed up the elimination of toxic chemicals, either exogenous or endogenous and terminate the exposure.

**Dileep K. Singh**
**All rights reserved - © 2012 Bentham Science Publishers**

All the organs have varying degree of exposure to the toxic chemicals. Therefore, even if metabolism occurs in all organs, the enzymes for metabolism are found to be relatively more active in some organs like liver, kidney and intestinal mucosa.

Inside the cell, bulks of the xenobiotic metabolizing enzymes are located in Endoplasmic reticulum. When tissue like liver are carefully homogenized, fragment of endoplasmic reticulum are converted into microsomes. Microsomes are exact replica of endoplasmic reticulum both morphologically and biochemically. Chemically it is composed of about 40% lipoprotein, 12% cellular protein and 50% ribonucleic acid (RNA). The enzymes located in the ER fraction of the cell are often referred as microsomal enzymes.

Microsomal enzymes such as Cytochrome P450 (CYP), NADPH Cyt c reductase and Flavin containing monooxygenase (FMO) play a very important role in the metabolism of pesticides. Besides microsomal enzymes, other enzymes located in the cytoplasmic fraction of cells like Dehydrogenase, Reductase and Glutathione s transferase are also involved in pesticide metabolism.

## 4.2. PHASE I METABOLISM

As mentioned earlier, the majority of the xenobiotic compounds entering the body are lipophilic in nature, a property that helps them to penetrate the lipid membrane easily. The first and most important step in the metabolism of any xenobiotic compound is to introduce a polar group into the molecules. It thus increases the solubility of the compound and renders it a suitable substrate for Phase II reactions.

There are three types of Phase I reaction; Oxidation, Reduction and Hydrolysis.

### 4.2.1. Oxidation Reactions

An oxidation reaction can be defined as a reaction in which an element lost an electron or a molecule increase the proportion of oxygen. In brief, it is an addition of oxygen into compound. It is the most important reaction in phase I metabolism. Oxidation of pesticide can be broadly classified into two groups,

1. Microsomal oxidation or Monooxygenation.

2. Extramicrosomal oxidation.

#### *4.2.1.1. Microsomal Oxidation Reactions*

Microsomal oxidation or Monooxygenation or mixed-function oxidations are those reactions in which one atom of molecular oxygen is incorporated into the substrate while the other atom is reduced to form water. The electron involved in reduction of cytochrome P 450 or FAD is derived from NADPH. It is catalyzed by groups of enzymes such as the Cytochrome P450 monooxygenase (CYPs) system or Flavin-containing monooxygenase (FMOs) system.

#### *4.2.1.1.1. Cytochrome P450 Monooxygenase System*

Cytochrome P-450 monooxygenase system consist of

- Two flavoprotein (dehydrogenases), *i.e.* NADPH cytochrome P450 reductase and NADH cytochrome $b_5$ reductase.

- Two hemoprotein, *i.e.* cytochrome P 450 and cytochrome $b_5$.

- Two pyridine nucleotide, *i.e.* NADH and NADPH.

These enzymes remain embedded in the phospholipids matrix of the endoplasmic reticulum or microsomes. The phospholipids facilitate the interaction between the two enzymes.

## 4.3. CYTOCHROME P450

Cytochrome P450s is a group of isoenzymes, all of which possess an iron protoporphyrin IX as the prosthetic group. All have one common reductase, NADPH cytochrme c reductase, whose concentration is $^1/_{10}{}^{th}$ to $^1/_{30}{}^{th}$ to that of cytochrome P450 (Fig. **1**).

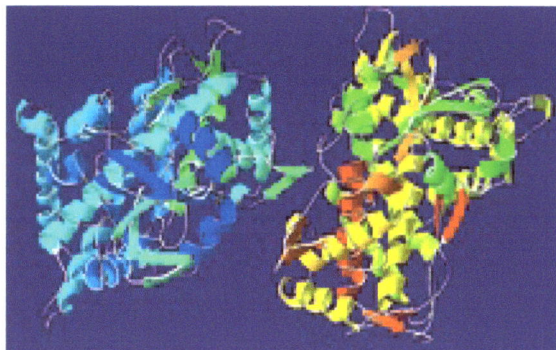

**Figure 1.** Cytochrome P 450 oxidase (CYP2C9) Source: http://en.wikipedia.org/wiki/Cytochrome P450

Cytochrome P450, a heme protein, which function as a terminal oxidase in steroid hydroxylation by bovine adrenal cortex microsomes was first reported by Estabrook et. al. (1963). It consists of a single polypeptide having a molecular weight of 45,000 daltons. It contains 1 mole of ferriprotoprophyrin IX, the iron which is bound to four pyrole nitrogens and two amino acid ligands, possibly cysteine and histidine.

There are more than 2000 cytochrome P450 types distributed throughout the living world. Cytochrome P450 is reported in microbes, yeast, insects and vertebrates. In vertebrates, it is mostly found in organs taking parts in xenobiotic metabolism *i.e.* liver, skin, gastrointestinal mucosa and even nasal mucosa. It thus reflects the body's defense mechanism against xenobiotic compounds at these portals of entry.

Unlike others cytochrome, Cytochrome P 450 derived its name not from the absorption maximum of the reduced form in the visible region of the spectrum, but from the unique wavelength of the absorption maximum of the carbon monoxide (CO) derivative of the reduced form, namely 450 nm.

## 4.4. NADPH CYTOCHROME C REDUCTASE

It is a stable flavoprotein having a molecular weight of about 78000 daltons and containing two mole of flavin adenosine diphosphate (FAD) per mole or one mole each of FAD and FMN (flavin mononucleotide) per mole of apoprotein. It is commonly analysed by the reduction of artificially added electron acceptors such as cytochrome c in the presence of NADPH. Hence, name NADPH cytochrome c reductase. It is found in the microsomal fraction of mammalian tissues involved in metabolism of xenobiotic compounds. It is also reported in insects and act as a mediator of electron flow from NADPH to cytochrome P450 or from NADH to cytochrome $b_5$.

One of the significant features of all microsomal enzymes is their inducibility by xenobiotic compounds. Thus stimulation of the metabolism of a chemical by prior administration of the same or another chemical is often taken as a presumptive evidence of its metabolism by the microsomal enzymes.

### 4.4.1.1.1. Mechanism of Cytochrome P450 Monooxygenation

The overall reaction of Cytochrome P450 monooxygenation is:

$$RH + O_2 + NADPH + H^+ \xrightarrow{\text{Oxidase}} ROH + H_2O + NADP^+$$

Where, RH is the substrate.

Here, only one atom of the $O_2$ molecule is inserted into the substrate, the other being reduced into water (Fig. **2**).

Cytochrome P450 mediated monooxygenation involves the following four basic steps:

1.  Binding of substrate.

2.  Reduction of the complex by NADPH cytochrome c reductase.

3.  Addition of $O_2$ and rearrangement of the complex.

4.  Reduction and loss of water molecule.

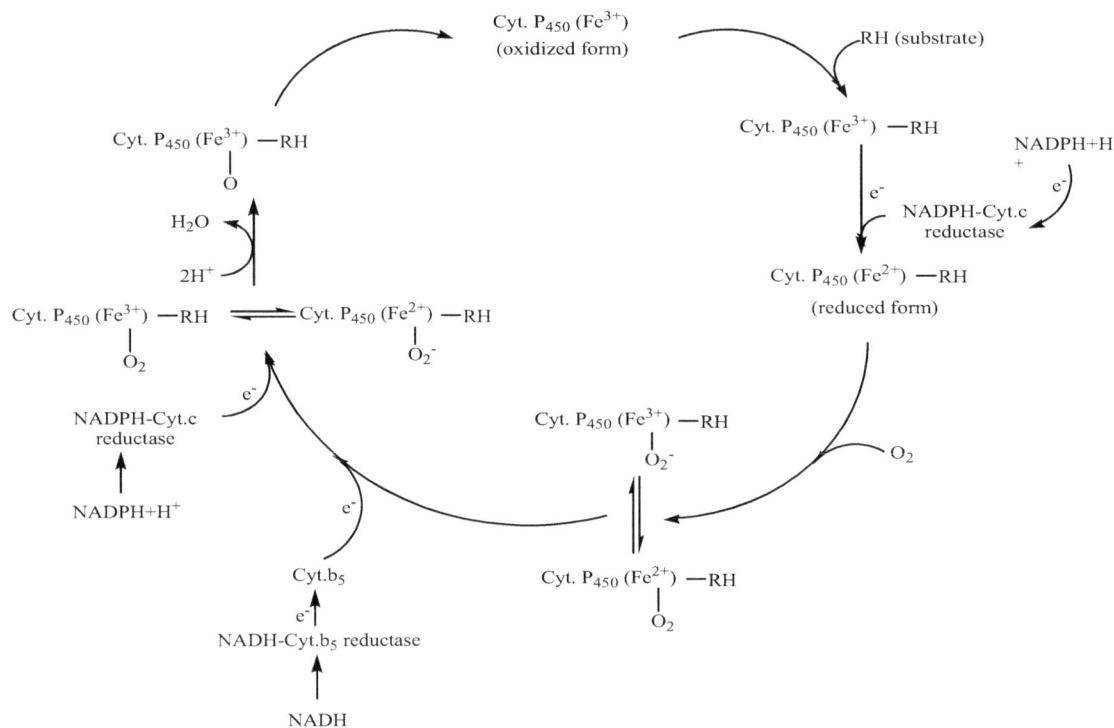

Schematic representation of Cytochrome P450 mediated monooxygenation reaction

**Figure 2.** Schematic representation of Cytochrome P450 mediated mono-oxygenation reaction.

### 4.5. SUBSTRATE BINDING

The binding of substrate take place when the cytochrome P450 is in the oxidized state *i.e.* $Fe^{3+}$, ferric state. Cytochrome P450 exist in two spin state- low spin, hexa coordinated and high spin, penta coordinated. Binding of substrate cause perturbation in the spin equilibrium of cytochrome P450, which lead conformational changes from low spin to high spin or vice versa. Such conformational changes give rise to particular spectral changes depending on the substrate.

Based on the particular range of spectrum, there are three types of substrate binding,

1.  Type I, the low spin conformation of cytochrome P450 change to high spin conformation after binding of a substrate. Example is binding of ethyl morphine.

2.  Type II, high spin to low spin conformational changes. Example is Aniline.

3.    Reverse type I or modified type II binding.

## 4.6. REDUCTION OF CYPS-SUBSTRATE COMPLEX BY NADPH CYTOCHROME C REDUCTASE

The iron atom of the heme group of cytochrome P450 is reduced from ferric, $Fe^{3+}$ to ferrous, $Fe^{2+}$. The reducing equivalent is transferred from NADPH through the enzyme NADPH cytochrome c reductase.

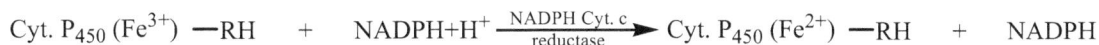

$$\text{Cyt. } P_{450} (Fe^{3+}) \text{—RH} \quad + \quad \text{NADPH+H}^+ \xrightarrow[\text{reductase}]{\text{NADPH Cyt. c}} \text{Cyt. } P_{450} (Fe^{2+}) \text{—RH} \quad + \quad \text{NADPH}$$

[Note: only one electron is transferred from NADPH to CYPs in this step]

## 4.7. ADDITION OF $O_2$ AND REARRANGEMENT OF THE COMPLEX

Reduced CYP-P450 substrate complex binds with molecular $O_2$ and undergo rearrangement. Though the oxidation state of oxygen and the iron in the ternary compound is not clear, but the oxygen may exist as a hydroperoxide or superoxide.

$$\text{Cyt. } P_{450} (Fe^{3+}) \text{—RH} \rightleftharpoons \text{Cyt. } P_{450} (Fe^{2+}) \text{—RH}$$
$$\underset{O_2^-}{|} \qquad\qquad\qquad \underset{O_2}{|}$$

$O_2$ hydroperoxide                                   $O_2$ superoxide

## 4.8. REDUCTION AND LOSS OF WATER

The CYP-substrate complex is oxidized by the second electron from NADPH is added through NADPH cytochrome c reductase, or alternatively by an electron donated by NADH through cytochrome $b_5$ reductase and cytochrome $b_5$.

The complex then rearrange with the loss of water and subsequent insertion of one atom of oxygen into the substrate. The oxidized substrate is then finally separated from CYP, leaving CYP in the oxidized state.

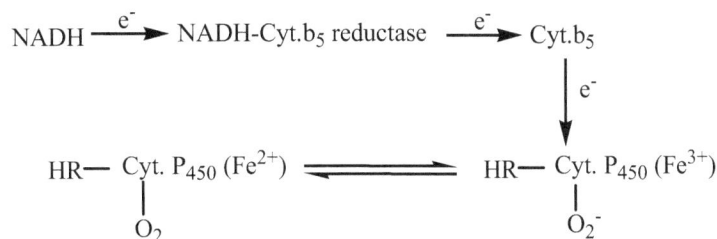

$$\text{NADH} \xrightarrow{e^-} \text{NADH-Cyt.}b_5 \text{ reductase} \xrightarrow{e^-} \text{Cyt.}b_5$$
$$\downarrow e^-$$
$$\text{HR—Cyt. } P_{450} (Fe^{2+}) \rightleftharpoons \text{HR—Cyt. } P_{450} (Fe^{3+})$$
$$\underset{O_2}{|} \qquad\qquad\qquad\qquad \underset{O_2^-}{|}$$

### 4.8.1.1.1. Cytochrome P450 Reactions

Microsomal monooxygenase enzymes are highly non-specific with the substrate as well as product belongs to wide range of chemical classes. Keeping this in mind that the same substrate may undergo more than one type of reaction, it will be more or less convenient to classify CYPs reactions on the basis of the type of chemical reaction.

Cytochrome P450 catalyzed reaction can be grouped into the following types,

1.    Epoxidation and Aromatic hydroxylation.

2.    aliphatic hydroxylation.

3.    N-dealkylation.

4.  O-dealkylation.

5.  S-dealkylation.

6.  S-oxidation.

7.  Ester cleavage and Desulfuration.

## 4.9. EPOXIDATION AND AROMATIC HYDROXYLATION

Epoxidation is an extremely important microsomal reaction because it not only leads to formation of stable and environmentally persistent epoxides, but also the arene oxides, an epoxide intermediate in hydroxylation of aromatic compounds (Figs. **3** & **4**). The enzyme responsible for epoxide formation is epoxidase.

The epoxidation of naphthalene was one of the earliest examples of an epoxide (arene oxide) as an intermediate in aromatic hydroxylation (Fig. **3**). The naphthalene epoxide thus formed can either rearrange non-enzymatically to yield predominantly 1-naphthol or interact with an enzyme, epoxide hydrolase to yield dihydrodiols. This shows that the aromatic hydroxylation generally (not always) proceeds through the formation of an epoxide intermediate.

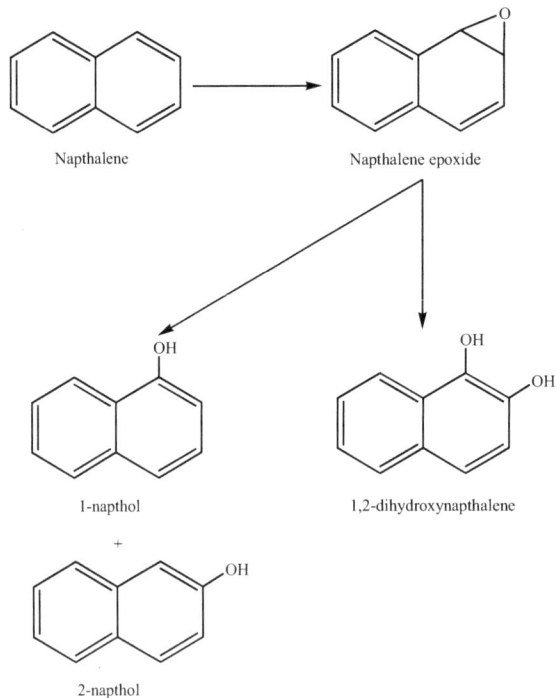

**Figure 3.** Metabolism of napthalene through epoxide formation followed by aromatic hydroxylation.

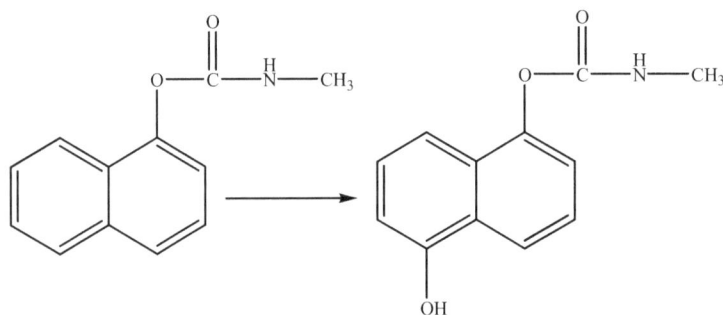

**Figure 4.** Aromatic hydroxylation of Carbaryl.

The phenomenon of intramolecular migration of the hydrogen atom on the aromatic ring of the compound during aromatic hydroxylation through epoxidation, is known as NIH shift (named after the National institute of Health, where it was discovered, Fig. **5**).

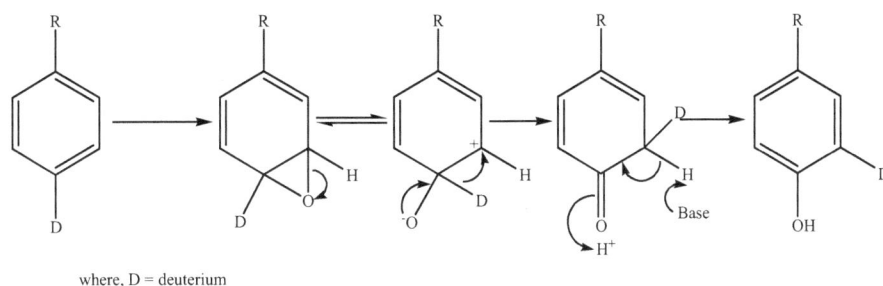

where, D = deuterium

**Figure 5.** Digramatic example of NIH shift.

Most cyclodiene pesticides are metabolized through epoxidation. The epoxidation generally occur at the double bond where the chlorine is absent (Fig. **6**).

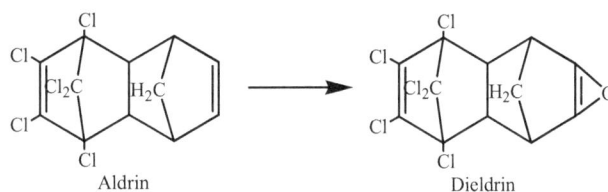

Aldrin                                    Dieldrin

**Figure 6.** Metabolism of Aldrin by epoxidation.

However, there are certain cases of epoxide formation is reported in saturated ring *i.e.* metabolism of Chlordane (Fig. **7**).

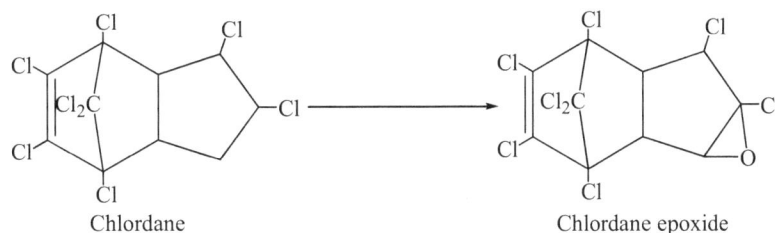

Chlordane                                    Chlordane epoxide

**Figure 7.** Epoxidation of Chlordane

Epoxides can undergo two different routes for excretion:

1. In compounds having 4a carbon atom, e.g. Aldrin and Dieldrin, a hydroxyl group get attached to the 4a carbon atom and go to phase II reaction where it will conjugate with Glucuronic acid and get excreted.

2. In compounds which lack 4a carbon atom, e.g. Heptachlor and Chlordane, highly polar trans-dihydrodiol are formed by enzyme epoxide hydrolase and excreted directly without forming any conjugants.

It is not clear whether epoxidation of cyclodiene compounds constitute an activation or detoxification process, since both epoxide and parents compound are equally toxic. However it was later shown that epoxidation fasten the effect of toxicity in case of Aldrin. In case of Chlordane, housefly rapidly degrades the epoxide product by hydroxylation. The hydroxylated product being non-toxic, it is probably the detoxification process.

## 4.10. ALIPHATIC HYDROXYLATION

Aliphatic compounds as well as alkyl side chains of aromatic compound are readily oxidized by microsomal oxidase. Such oxidation processes generally result in detoxification of the compound. Example is hydroxylation of DDT to yield Dicofol (Fig. **8**).

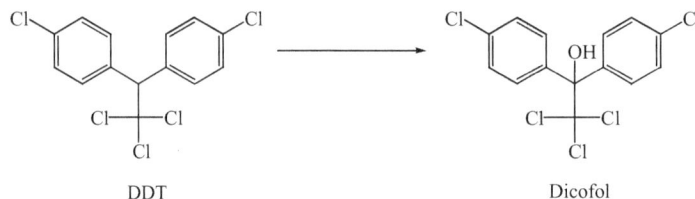

**Figure 8.** Aliphatic hydroxylation of DDT by microsomal oxidase.

## 4.11. N-DEALKYLATION

This refer to the reaction in which alkyl group attached to the electronegative nitrogen atoms or those in amines, carbamates or amides are removed oxidatively by conversion to the corresponding aldehyde. The reaction proceeds through an unstable α-hydroxy intermediate that spontaneously rearrange with the release of corresponding aldehyde (Figs. **9** & **10**).

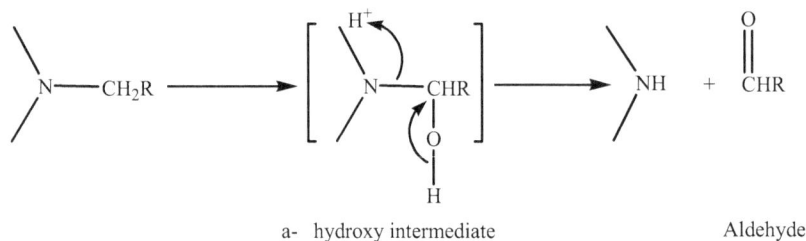

**Figure 9.** N-dealkylation reaction

The reaction generally results in the detoxification of the compound.

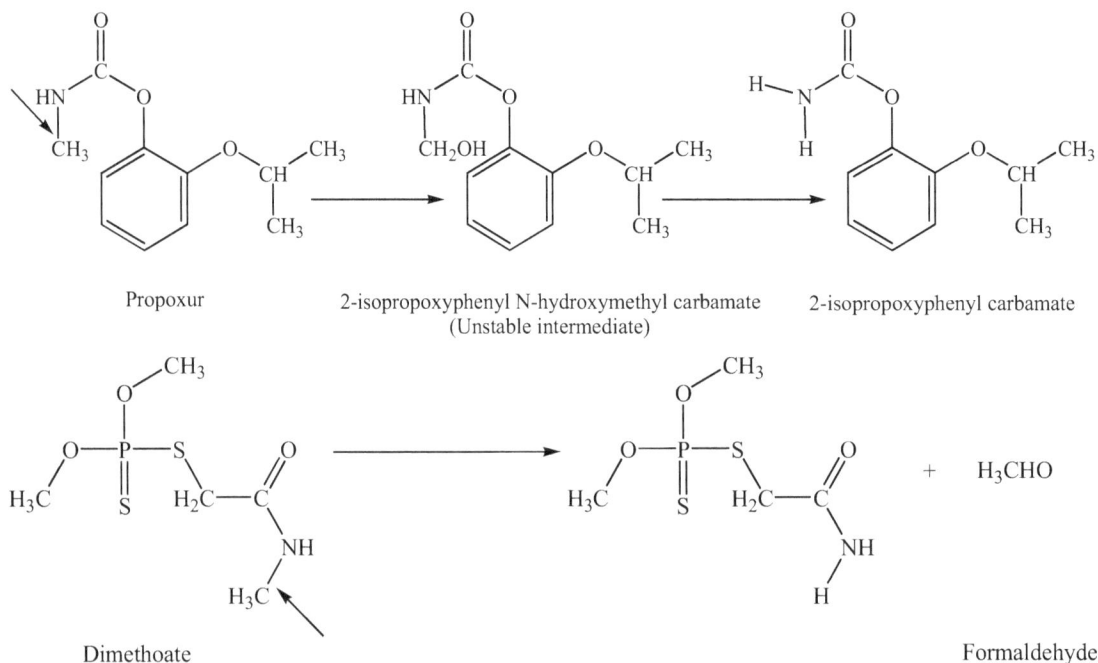

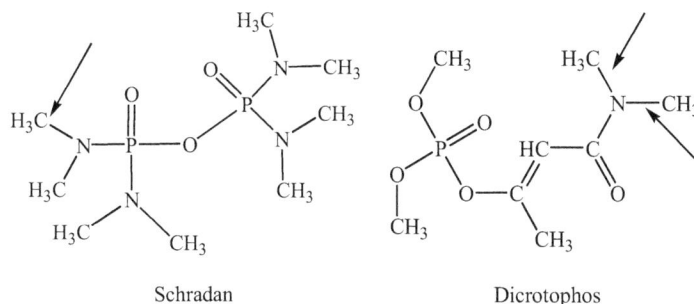

Schradan                    Dicrotophos

**Figure 10.** N-dealkylation reaction in different pesticides

## 4.12. O-DEALKYLATION

Ester and ether structure of insecticide readily undergoes O-dealkylation (Fig. **11**). However it involves formation of an unstable α-hydroxy intermediate as found in N-dealkylation. The α-hydroxy intermediate as in N-dealkylation, undergo spontaneous rearrangement to yield aldehyde in case of primary alkyl group or a ketone with a secondary alkyl group.

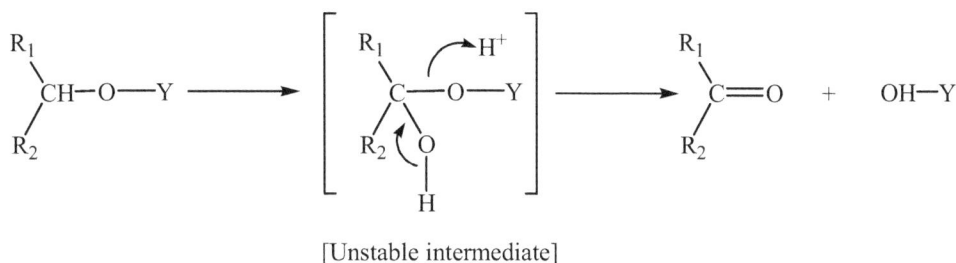

[Unstable intermediate]

Generalized reaction diagram O-dealkylation

**Figure 11.** Generalized reaction diagram of O-dealkylation

It is known to occur in wide range of Organophosphate insecticide, including certain dimethyl triester. It results in detoxification of the compound (Fig. **12**).

Parathion                              Phorate

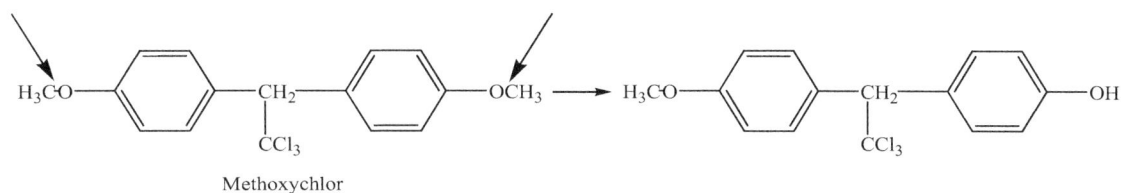

Methoxychlor

O-dealkylation of Parathion, Phorate and Methoxychlor

**Figure 12.** O-dealkylation of parathion, phorate and methoxychlor

## 4.13. S-DEALKYLATION

Microsomal enzyme system catalyzes S-dealkylation with oxidative removal of the alkyl group to yield corresponding aldehyde. However, it has been suggested that unlike N-dealkylation another enzyme system such as microsomal FAD-containing monooxygenase system may be involved.

Microsomal S-dealkylation is reported from the *in-vivo* metabolism of Aldicarb, with the conversion of methylthiocarbon to carbon-dioxide in housefly. However, S-dealkylation does not occur in *in-vitro* metabolism of Aldicarb in rat liver (Fig. **13**).

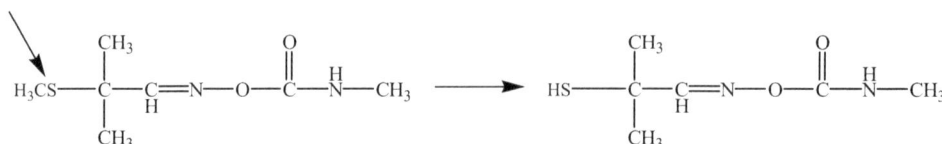

**Figure 13.** S-dealkylation of Aldicarb

## 4.14. DESULFURATION AND ESTER CLEAVAGE

Desulfuration reaction is also known as phosphorothioate oxidation. It is responsible for the increase insecticidal activity and mammalian toxicity of phosphorothionates $[(RO)_2P(S)OR']$ and phosphorodithiote $[(RO)_2P(S)SR']$, organophosphorus compounds. Because, the P=S group of these compounds are oxidatively desulfurated by microsomal monooxygenase to their respective P=O analogs, which is usually bind more tightly to acetylcholinesterase and are thus are more potent inhibitors of acetylcholinesterase. Much of the splitting of the phosphorus ester bonds in organophosphorus insecticides, formerly believed to be due to hydrolysis, is now known to be due to Oxidative dearylation (Fig. **14**).

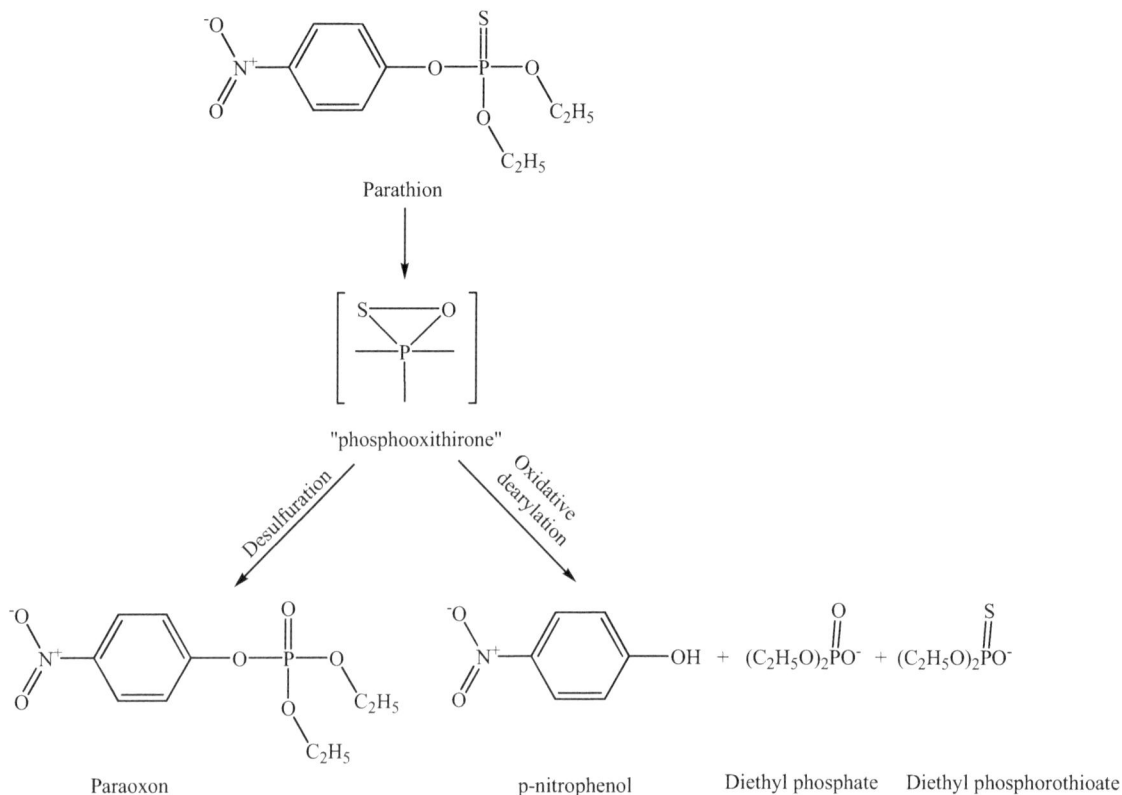

**Figure 14.** Desulfuration and oxidative dealkylation of Parathion

Oxidative dearylation as well as oxidative desulfuration involve a common intermediate of "phosphooxithirane" type.

## 4.15. S-OXIDATION

Thioethers compounds such as many organophosphates and carbamates compound are oxidized by microsomal monooxygenase to sulfoxide. Some of these sulfoxides are further oxidized to Sulfones. In general, sulfoxides formation represents an oxidative activation process leading to an increase in anticholinesterase activity (Fig. **15**).

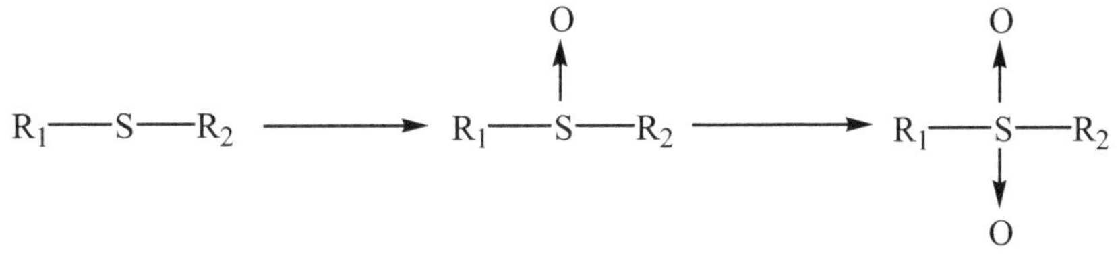

Sulfoxide                                      Sulfones

General reaction of S-oxidation

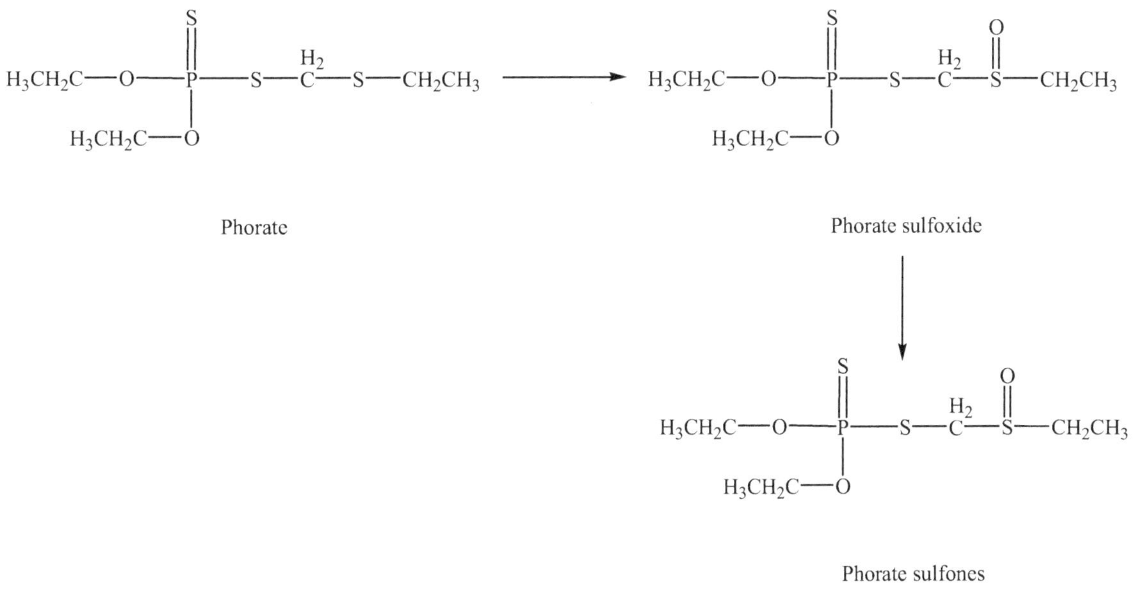

Phorate                                                  Phorate sulfoxide

Phorate sulfones

**Figure 15.** S-oxidation of Phorate

## 4.15.1.1.1. Flavin-contantaining Monooxygenase System

It is also one of the important microsomal enzymes, a monooxygenase requiring NADPH and oxygen, for the metabolism of pesticide. Earlier it was known as amine oxidase, but now it is known that it also catalyzed sulphur and phosphorus oxidation too. Unlike cytochrome P 450 monooxygenase system, Flavin containing Monooxygenase system catalyzed only oxidation and shows more or less substrate specificity (Fig. **16**).

The mechanisms of catalysis also differ slightly as the electrons are transferred directly from NADPH and not through NADPH cytochrome c reductase. Substrate containing soft nucleophiles e.g. Nitrogen, Sulphur, Phosphorus are good candidates for FMO oxidation. Examples include phorate, fonofos, methiocarb and many more. Oxidation by FMO generally results in detoxification of this substrate, however there are live cases of bioactivation in substrates involving sulphur oxidation.

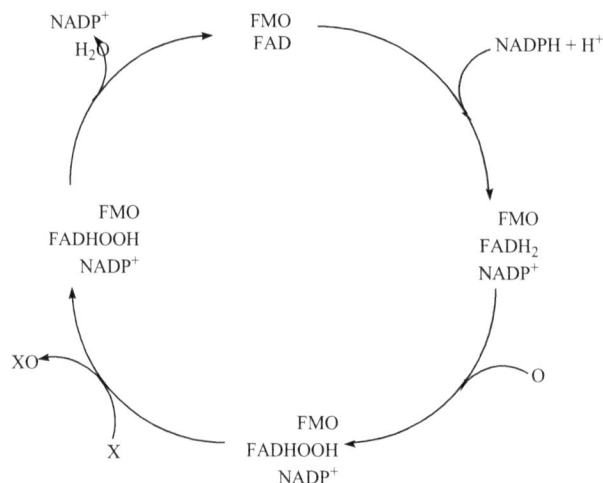

**Figure 16.** Reaction mechanism of Flavin-monooxygenase system (FMO).

Since both CYP and FMO require NADPH and oxygen, it is very difficult to distinguish which enzymes catalyzed an oxidation process. Above all, it is also known that most FMO substrates are also substrate for CYP (Fig. **17**).

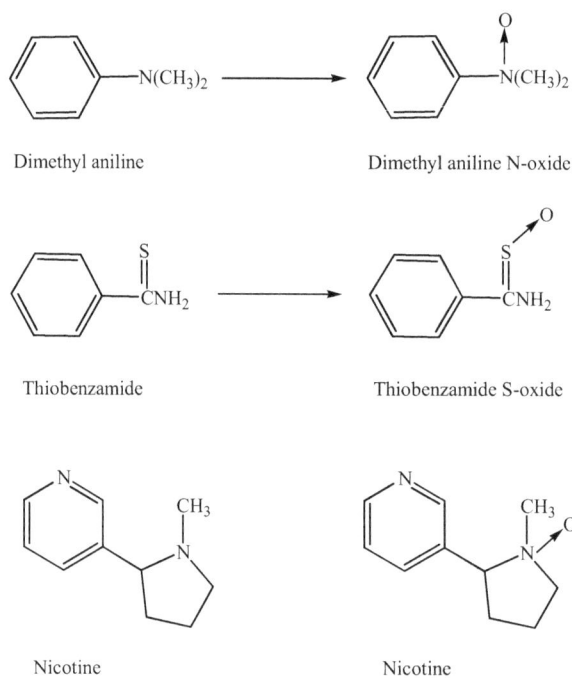

**Figure 17.** Examples of microsomal oxidation catalyzed by Flavin-containing monooxygenase.

Here the conversion of nicotine into cotinine *i.e.* it is catalyzed by two enzymes acting in sequences, CYP followed by soluble aldehyde dehydrogenes (Fig. **18**).

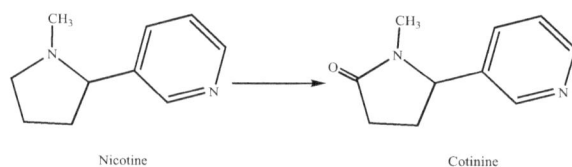

**Figure 18.** Nicotine degradation.

**Learning the facts :**

Metabolism of pesticide is conveniently classified under two headings:

- **Phase I reactions**: A functional (polar reactive) group is introduced into the molecule by oxidation, reduction or hydrolysis which produce materials for phase II reactions.

- **Phase II reactions** –These are conjugation reactions with an endogenous substrate and involve the covalent linkage of the toxin or phase I product to a polar compound.

    (i)   Conjugation processes usually required cooperation between membrane bound enzymes in the microsomes and other enzymes and cofactors present in the cytosol.

    (ii)  The phase II products are usually more water soluble than the parent compound and so are more readily excreted.

Microsomal oxidation/Monooxygenation of pesticides are catalyzed by either cytochrome P450 dependent momooxygenase system or FAD containing monooxygenase. Oxidation involves the enzymatic addition of oxygen or removal of hydrogen.

- Cytochrome P 450 (CYP 450) is very large and diverse superfamily of hemoproteins found in all living organisms. The active site of cytochrome P450 contain an iron atom in a porphyrin complex.

- The activity of cytochrome P 450 enzymes is called monooxygenase activity since one atom of molecular oxygen is used in formation of hydroxyl group and other is reduced by NADPH + $H^+$ to water. Under certain circumstances it may catalyze reduction reaction.

- The NADPH requiring general oxidation system, commonly referred to as the "Microsomal oxidase system" (also known as monooxygenase or mixed function oxidase system, MFO), is located in the microsomal portions of various tissues, particularly in the liver.

**Components of microsomal oxidases**

Mixed function oxidase system (MFO) is characterized by:

1.   Requiring NADPH as a cofactor,

2.   Involving an electron transport system with cytochrome P 450, and

3.   Being capable of oxidizing many different kinds of substrates (*i.e.* substrate nonspecificity).

The major components of the system which play the central role in oxidation are:

1.   A flavoprotein, NADPH-cytochrome c reductase, and

2.   A unique cytochrome, cytochrome P450

## NADPH cytochrome c reductase

1. NADPH cytochrome c reductase is recognized as a mediator of electron flow from NADPH to the oxygen activating enzyme.

2. It is stable flavoprotein, having molecular weight of 70,000 dalton and containing 2 moles of FAD per mole, and is commonly assayed by the reduction of artificially added electron acceptors such as cytochrome c or neotetrazolium in the presence of NADPH.

3. The level of NADPH cytochrome c reductase can be increased by the induction of microsomal oxidases and that antibody to NADPH cytochrome c reductase inhibits oxidative metabolism.

4. NADPH cytochrome c reductase has been found in the microsomal fraction of mammalian tissues such as liver, adrenal cortex, spleen, kidney, heart, and lung as well as in the tissues of insects.

## Cytochrome P 450

1. Cytochrome P 450 is carbon monoxide binding pigment of microsomes and are actually hemoprotein of b cytochrome type.

2. Cytochrome P 450 consists of a single polypeptide having a molecular weight of 45,000 daltons. It contains 1 mole of ferriprotoprophyrin IX, the iron of which is bound to four pyrrole nitrogens and two amino acid ligands, possibly cysteine and histidine.

3. The localization of this hemoprotein, however, is not totally restricted to the microsomal fraction, nor, for that matter, to any specific tissue or group of animals. Microbes and insects, in addition to mammals, are known to contain cytochrome P 450.

4. In mammals, it has been located in the microsomal fraction of extrahepatic organs such as the kidney, lung, and placenta and has also been reported in the mitrochondria of the adrenal cortex and corpus luteum.

5. Cytochrome P 450 is also found in the particulate fraction of yeast. In almost all vertebrates species which have liver, cytochrome P 450 is most important for oxidation of xenobiotic compounds.

6. It is also found in skin, nasal mucosa and gastrointestinal tract presumably reflecting defence mechanism at portal of entry.

7. Cytochrome P 450 is the common oxygen-activating enzyme for the entire family of microsomal mixed-function oxidases.

## Catalytic events of microsomal oxidase system

The process of microsomal oxidation integrates the transfer of electrons from NADPH with the binding of substrate and oxygen at cytochrome P450. Two separate one-electron reductions are involved *i.e.*

1. The first occurs after the initial complexing of the substrate with oxidized cytochrome P 450 and

2. The second on formation of the reduced cytochrome P-450/substrate/oxygen complex.

Subsequently to catalysis, oxidized cytochrome P-450 is regenerated by dissociation of the hydroxylated product and water. The mechanism of microsomal oxidation involves three basic events *i.e.*

1.  Substrate binding,

2.  Reduction, and

3.  Oxygen binding and activation.

### Substrate binding

Substrate binding is the initial reaction in the microsomal oxidase system in which electrons from NADPH integrates the binding of substrate and oxygen at cytochrome P-450.

### Reduction

The overall process of microsomal oxidation requires the transfer of two electrons from NADPH through a series of redox components to cytochrome P450. The expected 1:1 ratio for substrate oxidation and NADPH oxidation can be known from the general equation:

$$S + O_2 + XH_2 \longrightarrow SO + H_2O + X$$

Reduction occurs in two separate one-electron steps.

1.  Involving the reduction of the cytochrome P-450-substrate complex *via* NADPH-cytochrome c reductase is monitored in the presence of carbon monoxide by the appearance of absorbance at 450 nm.

2.  The second electron is introduced at the level of the oxygenated cytochrome P450-substrate complex (oxycytochrome P 450).

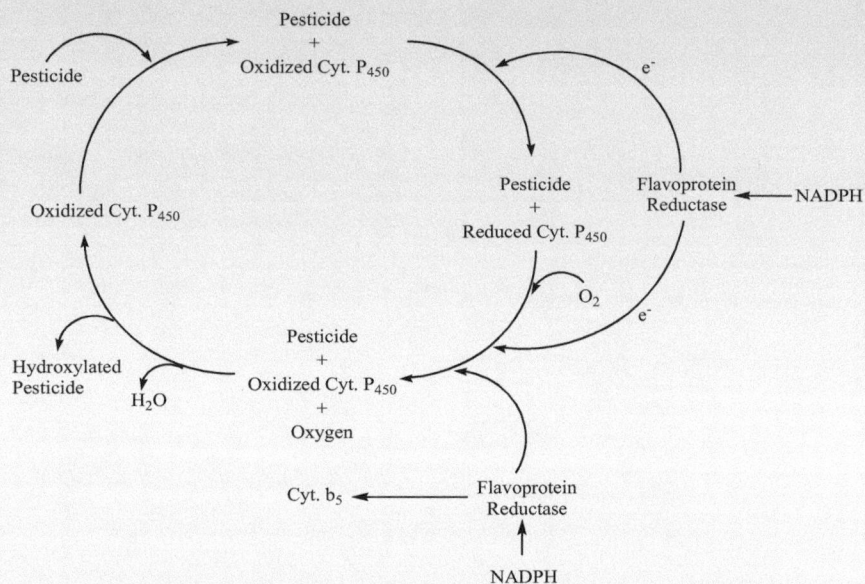

**Figure 19.** Interaction of electrons, oxygen, and substrate in the microsomal oxidase System.

The route of the second electron from a reduced pyridine nucleotide to oxy cytochrome P-450 reputedly involves cytochrome $b_5$, a component of the microsomal electron transport system usually associated with fatty acid desaturation.

## Oxygen binding and Activation

1.  Binding of carbon monoxide to cytochrome P 450 in competition with oxygen is an indication of the role of this cytochrome in oxygen activation (Fig. **19**).

2.  In fact, along with the requirements for NADPH and oxygen, inhibition by carbon monoxide is an important criterion for cytochrome P 450-mediated microsomal oxidation.

3.  Subsequent to its binding to cytochrome P 450, the oxygen molecule is activated and split, one atom being inserted into the substrate and the other reduced to water.

4.  The mechanism by which cytochrome P 450 effects the introduction of oxygen into the substrate could conceivably involve the generation of a free radical or the direct insertion of atomic or molecular oxygen into the substrate.

## Microsomal oxidation of pesticides

Most organic insecticides and synergists are subject to microsomal oxidase system. Many of them possess multiple sites at which oxidation can occur, and consequently a combination of several transformations can takes place with any particular compound (Fig. **20**).

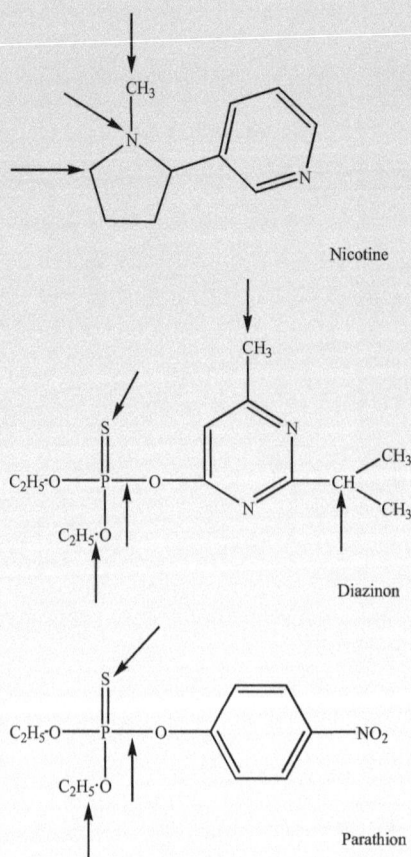

**Figure 20.** Examples of multiple sites for microsomal oxidation.

The reactions catalyzed by this system include,

1. O-, S-, and N- Alkyl Hydroxylation

2. Desulfuration

3. Epoxidation

4. Thio ester oxidation

5. Aromatic hydroxylation

## O-, N-, and S- Alkyl Hydroxylation

1. O-, N-, and S- Alkyl Hydroxylation is an important pathway of organophosphates metabolism in which an alkyl group adjacent to a hetero atom such as oxygen, sulfur, and or nitrogen is a potential target for microsomal hydroxylation, but because of the electronegativity of the hetero atom , the reaction often leads to dealkylation.

2. These reactions usually takes place in microsomes, however, these may also possibly occur in cytoplasm in minor quantity when specific enzymes are present.

3. Enzymes responsible for these reactions are dealkylase, desulfurase, and hydrolase.

## O- dealkylation

Dealkylation of O-alkyl groups of the ester or ether structures of insecticides occurs readily, but does not take place by simple replacement of an alkoxy group with a hydroxy group. Instead, an unstable a-hydroxyl intermediate is produced which spontaneously releases an aldehyde in the case of a primary alkyl group and a ketone in the case of a secondary alkyl group (Figs. **21** & **22**).

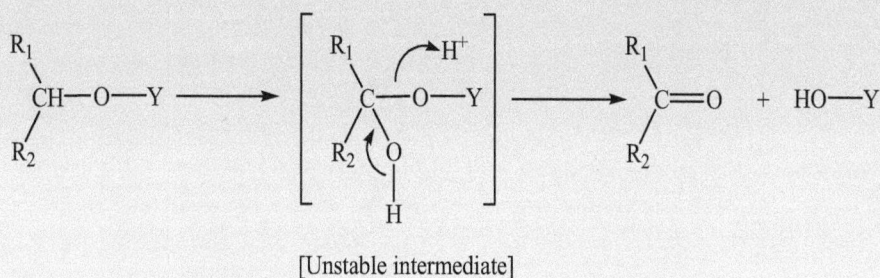

[Unstable intermediate]

Generalized reaction diagram O-dealkylation

**Figure 21.**General reaction for O-dealkylation.

## S- dealkylation

Microsomal S- demethylation of several methylthio compounds has been reported (Fig. **23**), and its involvement in aldicarb metabolism has been inferred from the *in vivo* conversion of the methylthio carbon to carbondioxide in the housefly. However, no S- demethylation of aldicarb has been detected in the rat liver *in vitro* system.

## O-Dealkylation

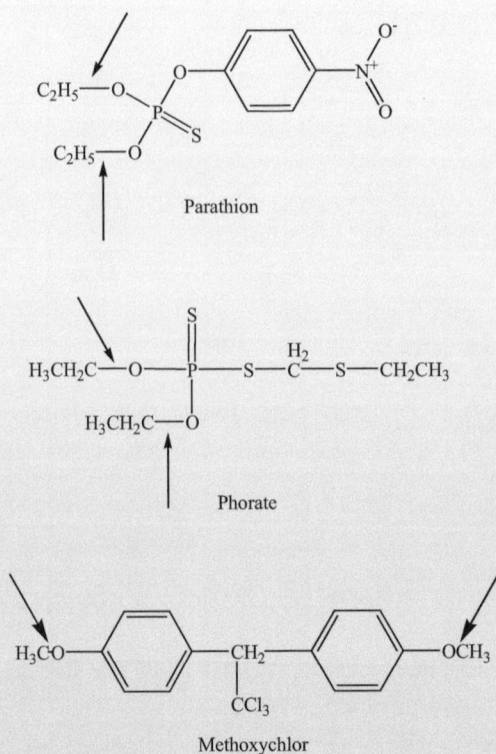

Parathion

Phorate

Methoxychlor

**Figure 22.** O-dealkylation reactions in different insecticides.

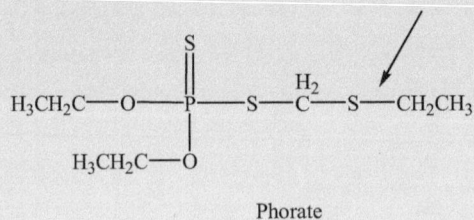

Phorate

**Figure 23.** S-dealkylation of Phorate.

## N- dealkylation

1.  N-dealkylation occurs in the metabolism of many organophosphates and carbamates. Unlike *O- dealkylation*, this reaction often yields a fairly stable N-a-hydroxy alkyl derivative, probably because nitrogen is less electronegative than oxygen.The metabolite may then undergo nonoxidative cleavage to a dealkylated product and an aldehyde.

2.  In the case of N-methyl hydroxylation, further oxidation to an N-formyl derivative has sometimes been noted, although the nature of the oxidase for this step has not been defined.

3.  The activation of the phosphoramidate insecticide, schradan to N-hydroxymethyl schradan is a classical example of N-dealkylation (Fig. **24**). It is likely that the N-hydroxymethyl derivative is an intermediate in stepwise N-demethylation reactions of this compound.

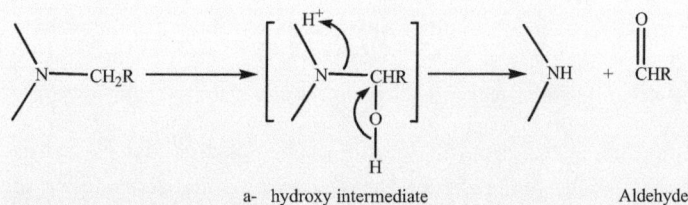

**Figure 24.** General reaction for N-dealkylation

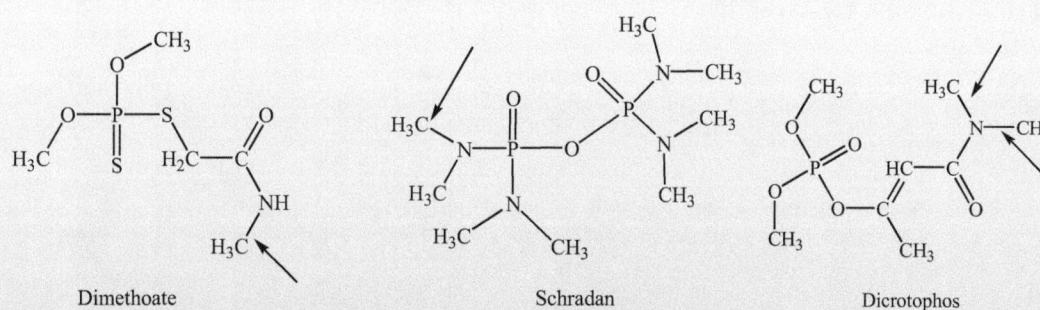

**Figure 25.** N-dealkylation reactions in different insecticides

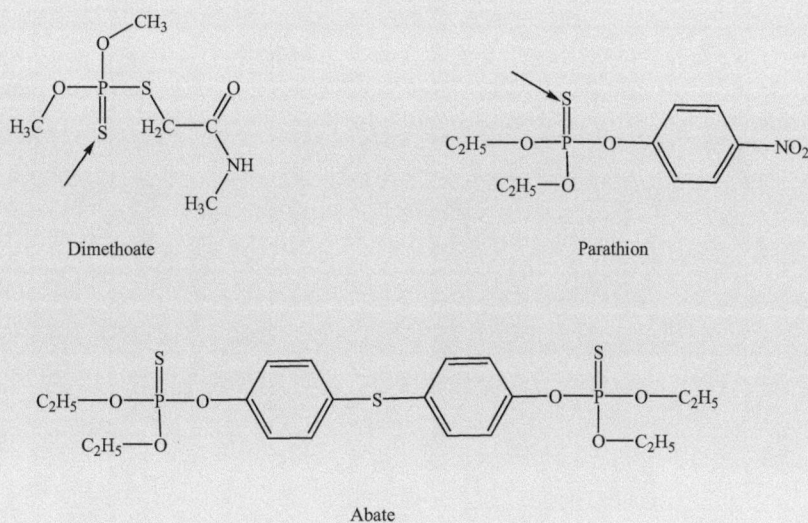

**Figure 26.** Desulfuration reactions in different insecticides

## Desulfuration

1. Desulfuration is one of the most commonest metabolic pathway of organophosphorus pesticides which contains phosphorothioate and phosphorodithioate esters (Fig. **26**).

2. P S structure of organophosphorus pesticides are desulfurated to their corresponding P O analogues by microsomal oxidases of mammals and insects.

3. The detached sulfur is apparently bound covalently to microsomal macromolecules and is eventually excreted as inorganic sulfate.

4. Desulfuration, however, represents only part of the microsomal oxidation of these compounds. These esters are concurrently hydrolysed to P S and the corresponding leaving groups by the oxidase system.

5.  Oxidative desulfuration of **P=S** to **P=O** is always lead to more toxic products. These changes are responsible for increase in their toxicity towards the cholinesterase enzyme where phosphorothioate activation to phosphate, an important and direct cholinesterase inhibitor.

6.  This type of oxons are reported in parathion, dimethioate, abate *etc.* The formation of oxons require NADPH and molecular oxygen with microsomes. The eliminated S is absorbed in microsomes and excreted in urine.

## Thioether or Sulfur Oxidation

1.  Sulfoxidation of phorate, aldicarb takes place in microsomes where one atom of oxygen is attached with S, forming sulfuroxide and when two atoms of oxygen are attached with S, forming sulfone (Fig. **27**).

2.  Enzyme responsible in this reaction is called sulfoxidase, they are named after the pesticide name, such as phorate sulphoxidase *etc.* These are soluble in organic solvent and responsible for mixed function oxidase reaction.

3.  Usually the alkyl sulfur in pesticide is rapidly oxidized to sulfoxide and more slowly to sulfones.

4.  This type of reaction is very limited in insecticide metabolism. Recently, it was observed that these sulfoxides enter into phase II reaction, where they conjugate with glutathione and finally become highly polar and excrete from body.

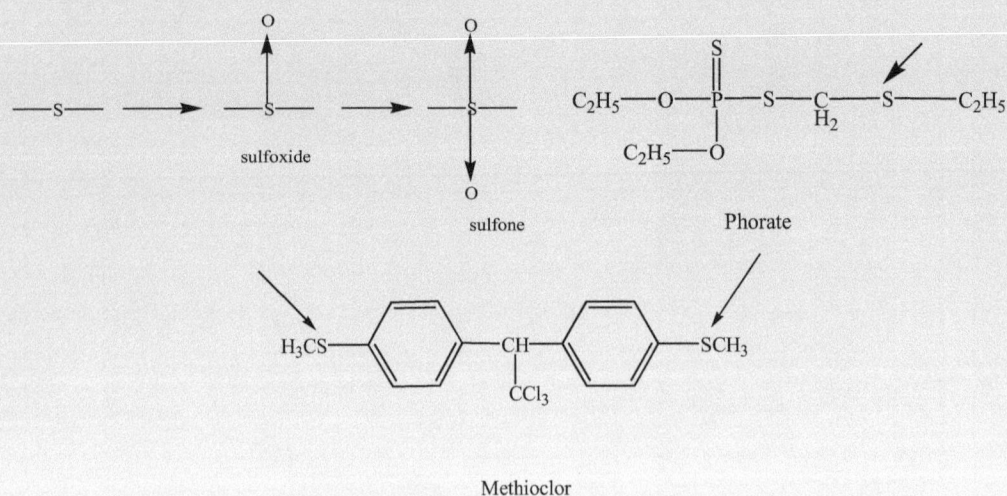

**Figure 27.** Thioether or sulphur oxidation reactions

## Aromatic hydroxylation or NIF shift

1.  The NIF shift (named after the National Institute of Health, where it was discovered) is a characteristic of aromatic hydroxylation by all mixed function oxidases (Fig. **28**).

2.  During such hydroxylation reactions, the hydrogen atom replaced by the hydroxyl group is not always expelled from the molecule, but may migrate to an adjacent position in the ring.

3.  The degree of hydrogen retention varies with different substrates. Substitution other than hydrogen (*e.g.,* halogen) may behave in an similar manner. Such retention is not observed in the well-known electrophilic substitiontion reactions.

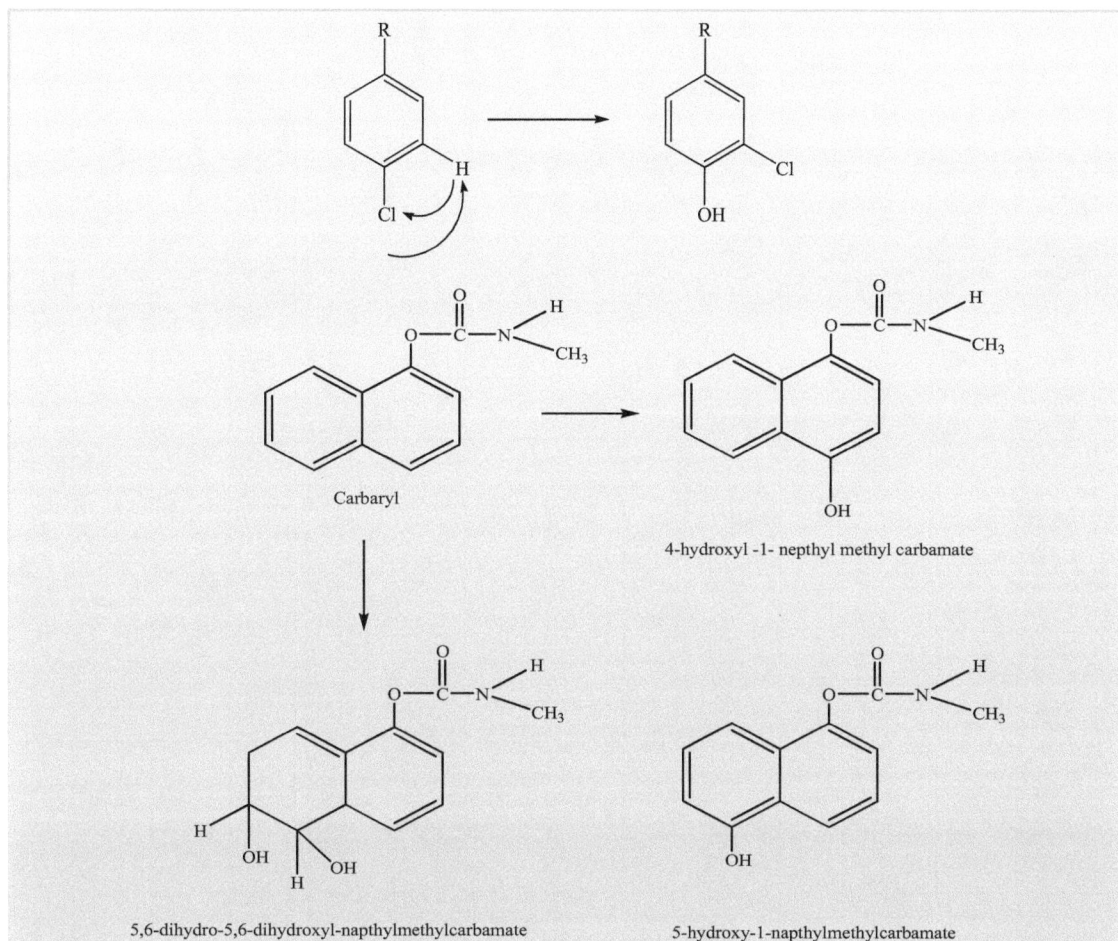

Figure 28. Aromatic hydroxylation or NIH shift reactions.

## EPOXIDATION

1. Epoxidation is an important microsomal reaction in which stable and environmentally persistant epoxides of dihydrodiols are formed (Fig. **29**).

2. It is one of the important pesticide degradation reaction in case of cyclodiene compounds e.g., heptachlor, aldrin, isodrin, when this reaction occurs, the oxygen gets detached to the place where chlorine is absent, but double bond is present.

3. These epoxides may further go for hydroxylation reaction and form trans-diols.

4. Enzymes responsible for this reaction is epoxidase which is present in microsomes. However, it is also reported from cytoplasm.

5. These epoxides are quite stable metabolites and responsible for pollution in environment.

6. Epoxides are subjected to hydration to form dihydrodiols, by epoxide hydrases.

7. Trans hydrodiol are also catalysed by other enzyme epoxide isomerase. From here, they move into phase II reaction, where they conjugate with glucoronic acid and become highly polar and excreted from body.

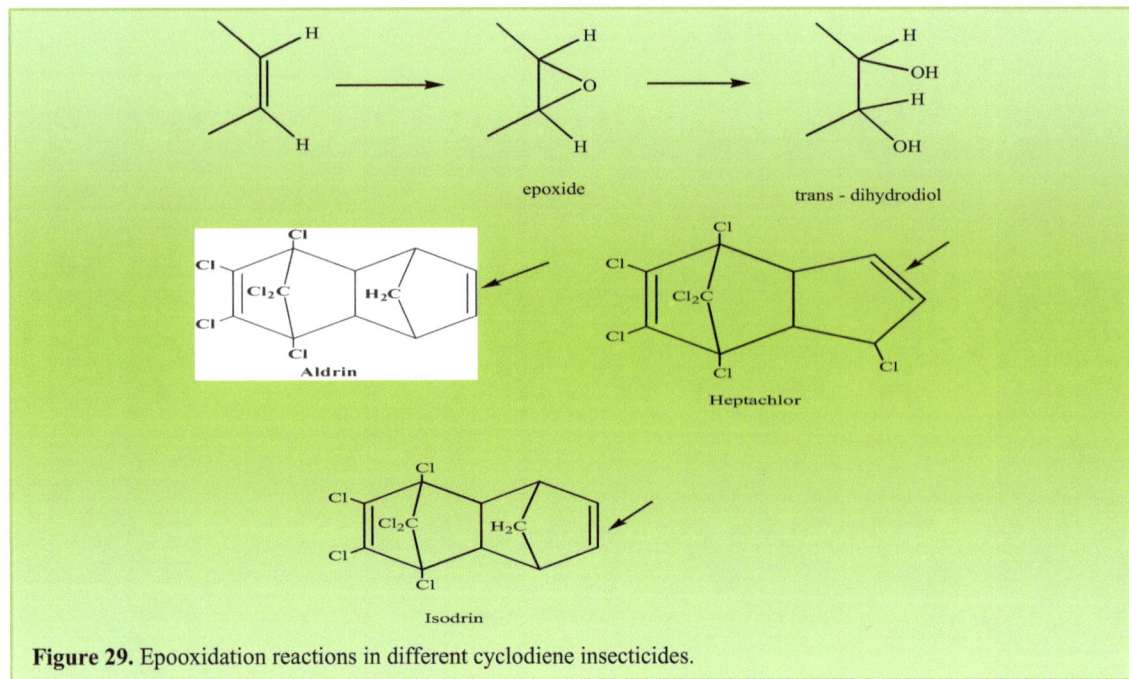

**Figure 29.** Epooxidation reactions in different cyclodiene insecticides.

## 4.16. EXTRAMICROSOMAL METABOLISM OF PESTICIDES

Besides the two microsomal monooxygenase system, other non-microsomal enzymes are also involved in the oxidation of xenobiotic compound. The enzymes are located in the mitochondria or the soluble cytoplasm of the cell. They are either flavoprotein (e.g. monoamine oxidase in mitochondria) or pyridine nucleotide-linked dehydrogenase (e.g. alcohol and aldehyde dehydrogenase in cytoplasm)

### 4.16.1. Alcohol Dehydrogenase

It catalyzed the conversion of alcohol to aldehyde or ketones.

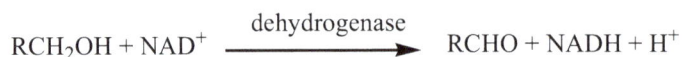

$$RCH_2OH + NAD^+ \xrightarrow{\text{dehydrogenase}} RCHO + NADH + H^+$$

Unlike monooxygenation of ethanol by microsomal CYP, this alcohol dehydrogenase reaction is reversible.

### 4.16.2. Aldehyde Dehydrogenase

Aldehydes being highly reactive electrophilic compounds can cause various harmful effects to the body. Aldehyde dehydrogenase hydrolyzed aldehyde through various metabolic processes into carboxylic acids.

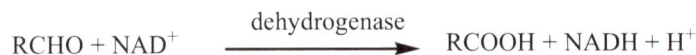

$$RCHO + NAD^+ \xrightarrow{\text{dehydrogenase}} RCOOH + NADH + H^+$$

These acids are then available as substrate for conjugation in the phase II reaction.

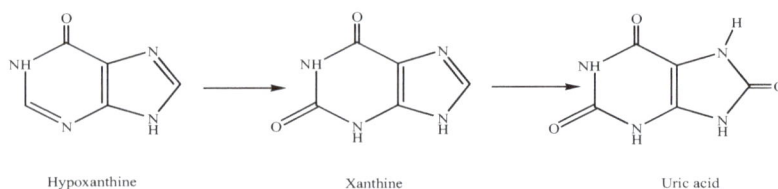

Hypoxanthine                    Xanthine                    Uric acid

**Figure 30.** Oxidation of hypoxarthine to xanthrine oxidase

Other enzymes which have similar function are aldehyde oxidase and xanthine oxidase, both are flavoproteins which contain molybdenum (Fig. **30**).

## 4.16.3. Reduction

Pesticides which contain functional group like nitro-, diazo-, disulfide, sulfoxide and alkene are susceptible to reduction. The enzymes reductase may be located in both microsomal fraction as well as soluble cell fraction. Cytochrome P450 and NADPH cytochrome c reductase are also capable of reducing nitro and azo group, in condition of low oxygen concentration. However oxygen will inhibits such reactions.

In insects, three types of reduction reaction are known to occur- nitro reduction, azo reduction and aldehyde or ketone reduction.

### *4.16.3.1. Nitro Reduction*

In bacterial and mammalian nitroreductase system, the reaction is catalyzed by CYP, hence are inhibited by oxygen. The nitro group of aliphatic as well as aromatic compound is reduced to amines (Fig. **31**).

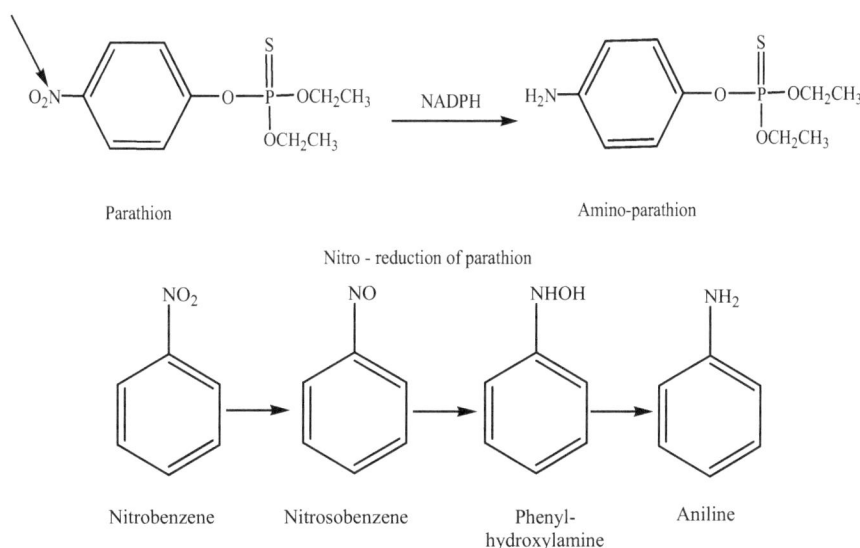

Parathion                                                                    Amino-parathion

Nitro - reduction of parathion

Nitrobenzene          Nitrosobenzene          Phenyl-          Aniline
                                              hydroxylamine

**Figure 31.** Nitro-reduction of Nitrobenzene.

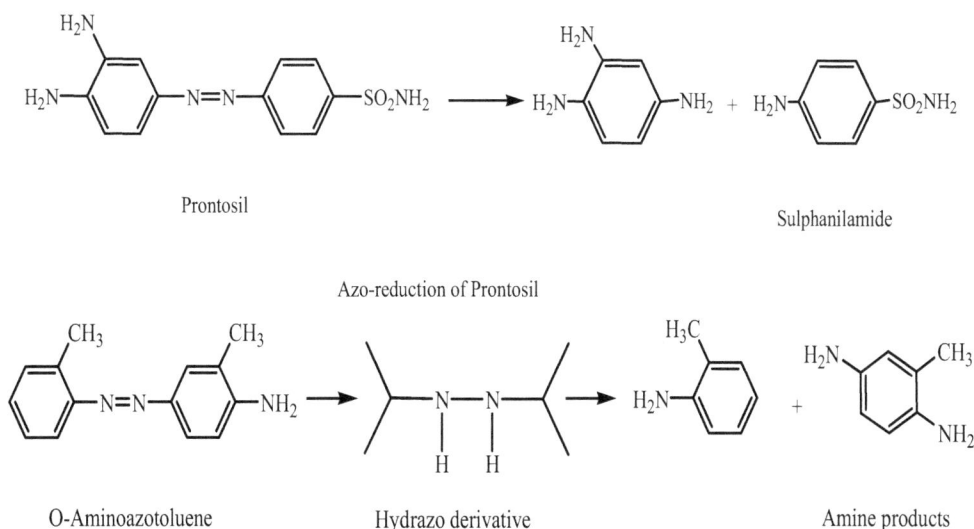

Prontosil                                                Sulphanilamide

Azo-reduction of Prontosil

O-Aminoazotoluene          Hydrazo derivative                    Amine products

**Figure 32.** Azo-reduction of O-Aminoazotoluene

### 4.16.3.2. Azo Reduction

Azo reduction also requires both anaerobic condition and NADPH, like nitro-reduction. It is inhibited by carbon monoxide and probably involves CYP (Fig. **32**).

### 4.16.3.3. Ketone and Aldehyde reduction

Besides the reverse reaction of alcohol dehydrogenase catalyzed reduction reaction, aldehyde reductase too is capable of reducing aldehyde and ketone into their corresponding alcohol group. Aldehyde reductase is cytoplasmic enzymes of low molecular weight. They require NADPH as cofactor (Fig. **33**).

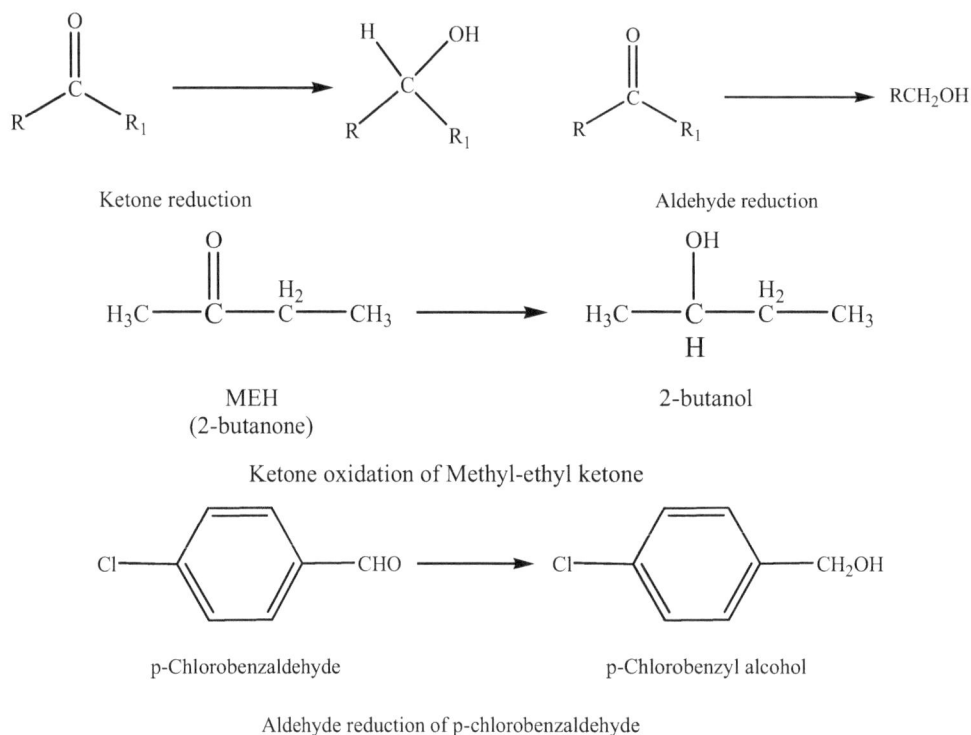

Ketone reduction                                        Aldehyde reduction

MEH
(2-butanone)                                             2-butanol

Ketone oxidation of Methyl-ethyl ketone

p-Chlorobenzaldehyde                                    p-Chlorobenzyl alcohol

Aldehyde reduction of p-chlorobenzaldehyde

**Figure 33.** Ketone and Aldehyde reduction.

### 4.16.3.4. Reductive dehalogenation and dehydrohelogenation

Reductive dehydrohelogenation (dehydrochlorination) of DDT to DDE (TDE) by anaerobic liver in the presence of NADPH is a well known reaction in the metabolism of DDT. However it is not clear whether this reaction proceeds enzymatically or non-enzymatically (Fig. **34**).

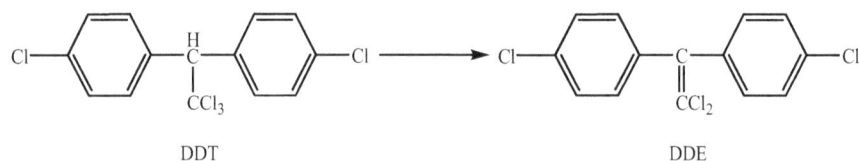

DDT                                                      DDE

**Figure 34.** Reductive dehydrochlorination of DDT

Dehydrohalalogenation reaction involve the removal of HX (X= halogen) from the substrate. The enzyme responsible for this reaction is dehydrohelogenase, a glutathione dependent enzyme. Here, the HCl is removed from DDT molecule.

In dechlorination, the DDT degrades into DDD by removal of one chlorine atom and addition of hydrogen atom.

## 4.16.4. Hydrolysis

Insecticides, especially organophosphate compounds, which possess ester linkage, are subjected to hydrolysis by enzyme esterase.

Esterases are hydrolases that split ester compound by addition of $H_2O$ to yield an acid and an alcohol.

$$RCOOR_1 \quad + \quad H_2O \longrightarrow RCOOH \quad + \quad R_1OH$$

Ester                         Acid         Alcohol

On the basis of the susceptibility and metabolism of organophosphate compounds, there are three types of esterase:

1.  A-esterase or aryl-esterase, they are not inhibited by organophosphate compounds and hydrolyzed them.

2.  B-esterase, it is the largest and most important group of esterase. They are susceptible to organophosphate inhibition. The inhibition is due to phosphorylation of the serine residue present in their active site.

3.  C-esterase or acetyl-esterase, they are not susceptible to organophosphate inhibition, neither metabolized them. Usually they prefer acetyl ester group as substrate.

There are two groups of enzymes which are mainly important in the metabolism of insecticides:

1.  Carboxylesterase: It belongs to B-esterase group. They are responsible for the hydrolysis of organophosphates, carbamate, pyrethroid, and some juvenoids in insects.

They are responsible for the selective toxicity of Malathion that favors mammals over insects (Fig. **35**).

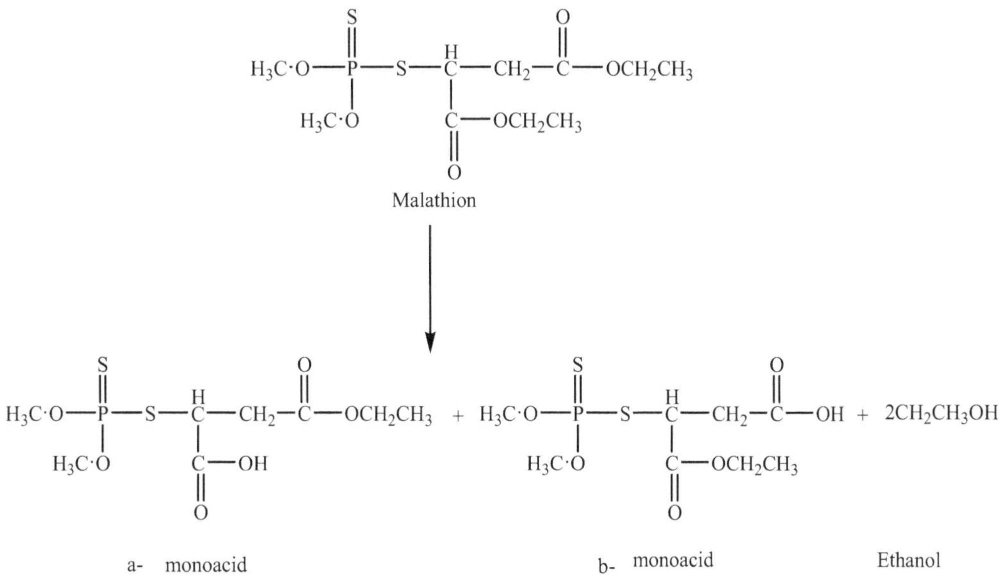

**Figure 35.** Hydrolysis of Malathion by Carboxylesterases.

1.  Phosphatase: it belongs to A-esterase or arylesterase group. They are responsible for hydrolysis of phosphate ester of many Organophosphate compounds.

Example: Hydrolysis of Paraoxon to diethyl phosphoric acid and p-nitro phenols (Fig. **36**).

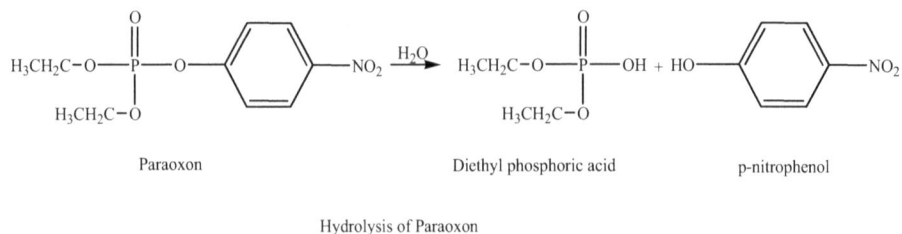

Hydrolysis of Paraoxon

**Figure 36.** Hydrolysis of Paraoxons.

Besides these, several amide containing organophosphate are hydrolyzed by carboxyamidase to their respective carboxylic acid derivatives (Fig. **37**).

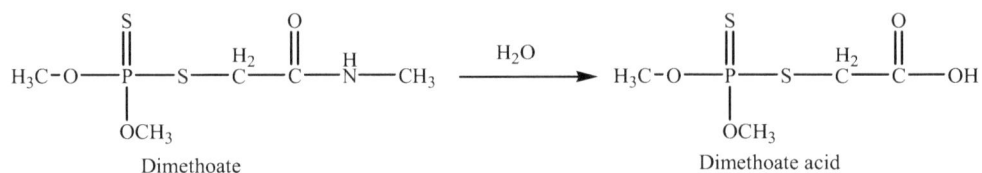

**Figure 37.** Hydrolysis of Dimethoate by Carboxyamidase.

### 4.16.5. Phosphotriesterase Hydrolysis

1. Degradation of organophosphorus pesticides by Phosphotriesterase is an important mechanism for pesticide detoxification outside of the microsomes, where enzyme attack in the phosphorus ester, an anhydride bond hence enzyme term as Phosphotriesterase (Fig. **37**).

There are 3 possible reactions in cytoplasm for detoxification-

- $1^{st}$ reaction lead to the formation of dialkyl phosphorothioic acid

- $2^{nd}$ reaction lead to the formation of dialkyl phosphoric acid

- $3^{rd}$ reaction leads to the formation to the metabolite desalkyl derivative and an alcohol. This type of reaction takes place in mammalian kidney and liver.

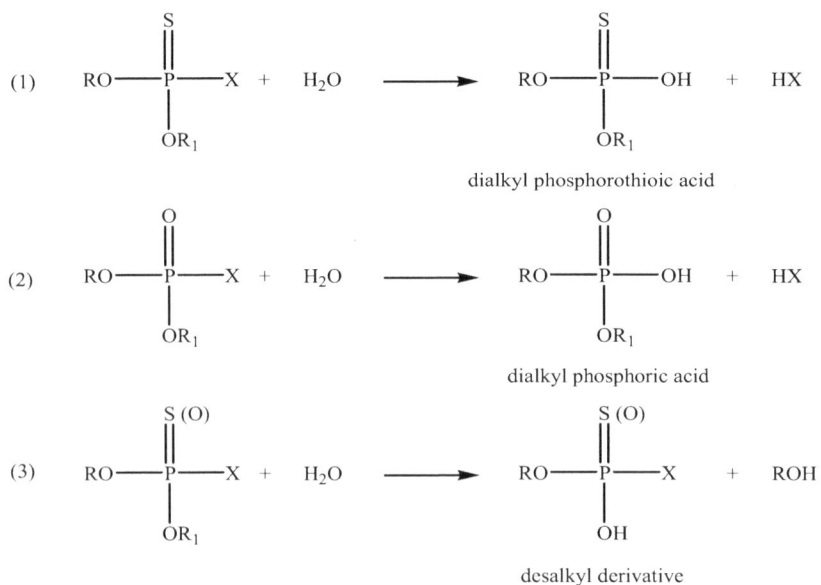

**Figure 38.** Phosphotriester hydrolysis in organophosphorus pesticide.

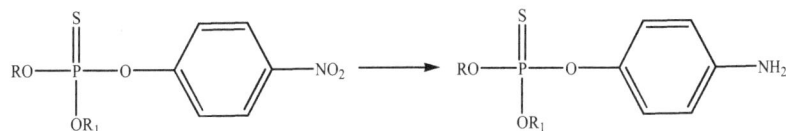

**Figure 39.** Nitroreductase reaction in organophosphorus pesticide.

### 4.16.6. Nitroreductase

The reduction of a number of nitro-containing organophosphorus compounds such as parathion, sumithion, and EPN is one of the minor detoxification reactions. The reduction of the nitro group occurs through the enzyme nitroreductases (Fig. **39**). Through this reaction, parathion is hydrolyzed *in vivo* to p-aminophenol, conjugated with glucoronic acid to an appreciable extent and become highly polar which excrete finally in the urine as p-aminophenyl glucoronide (Fig. **40**).

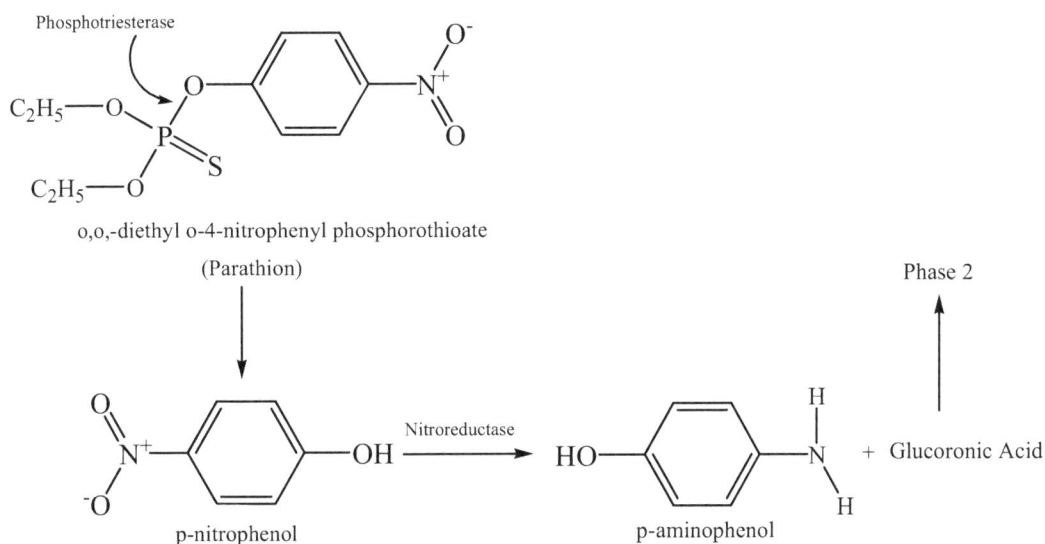

**Figure 40.** Nitroreductase reaction in parathion.

This type of enzymatic activity is reported in mammalian kidney, spleen, lungs, and erythrocytes, and also in avian kidneys. NADPH and a high concentration of FAD (1.23 moles) are necessary as cofactors but this reaction is not affected by the presence or absence of oxygen.

### 4.16.7. Carboxyl Ester Hydrolysis (Organophosphorus Compounds)

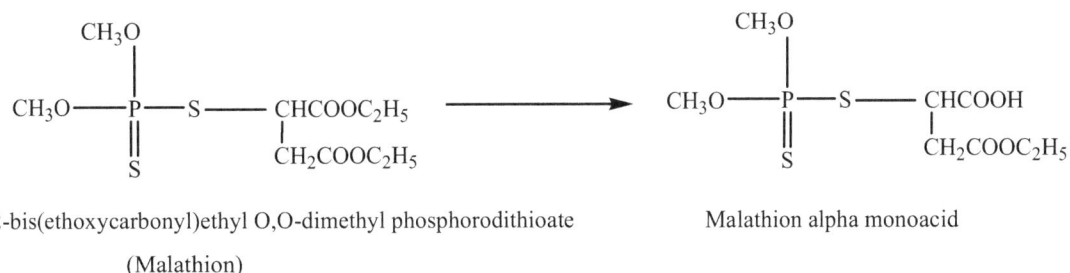

S-1,2-bis(ethoxycarbonyl)ethyl O,O-dimethyl phosphorodithioate                    Malathion alpha monoacid

(Malathion)

**Figure 41.** Carboxyl ester hydrolysis reaction in Malathion

The hydrolysis of both aromatic and aliphatic esters is catalyzed by carboxyl esterase, β-esterase, or aliesterase. But these enzymes are not responsible for the hydrolysis of choline esters (Fig. **41**). Carboxylesterases are very important in the metabolism of a number of types of insecticides, however, most importantly concern with the organophosphorus insecticides, Malathion, Acethion, *etc.* The hydrolysis of Malathion and acethion by this said enzyme involves cleavage of the carboxyester group to form a water-soluble nontoxic product. In this reaction Malathion and acethion forms the metabolite malathion monoacid and acethion monoacid respectively. Only one carbethoxy group of Malathion was found to be hydrolysed by carboxylesterase and formed α-monoacid. This enzyme is found to be widely distributed in mammalian tissues e.g. kidney, lung, spleen, *etc.* The said enzyme is also reported in microorganisms.

### 4.16.8. Carboxyl ester Hydrolysis (Pyrethroids)

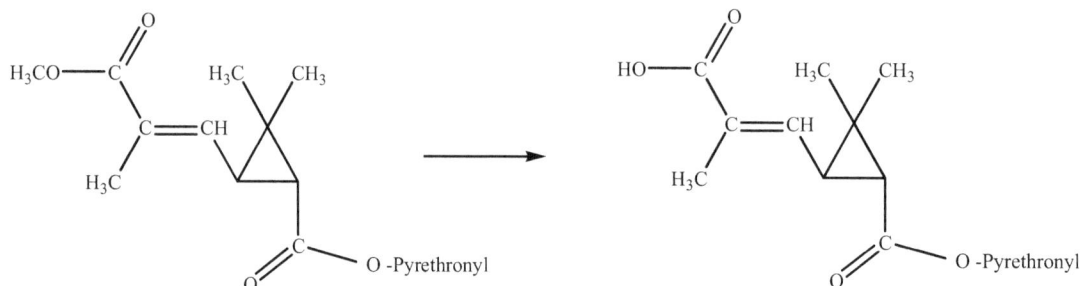

**Figure 42.** Carbosxyl ester hydrolysis in pyrethroids.

Carboxyl ester hydrolysis (Pyrethroids) is very much similar to that of organophosphorus (Fig. **42**). Compounds where ester linkage of compound allethrin, pyrethrum is the major target site of hydrolysis. Enzymes responsible are carboxyl esterases, β-esterases, and aliesterases, very much similar with that of organophosphorus compounds. These enzymes are reported in kidney, lung and spleen *etc.*

### 4.16.9. Carboxylamide Hydrolysis

The number of amide containing organophosphorus pesticide have been reported to metabolise by carboxylamide hydrolysis to their corresponding carboxylic acid derivative e.g. dimethoate, in this case, the product form is monocarboxylic acid which is the direct result of amidase action (Fig. **42**). The dimethoate monoacid was initially isolated from the urine however it was also reported in various vertebrate tissues.

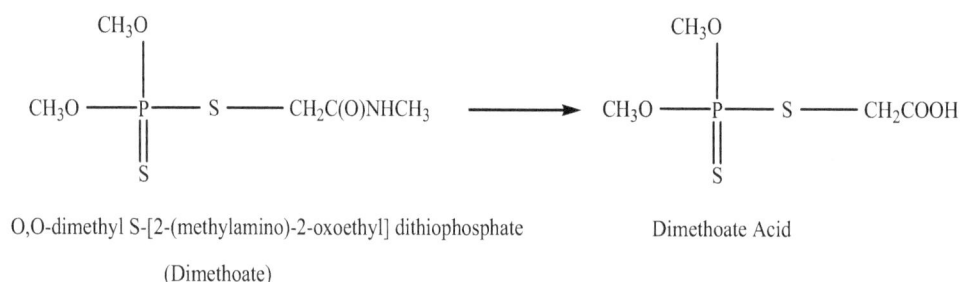

O,O-dimethyl S-[2-(methylamino)-2-oxoethyl] dithiophosphate

(Dimethoate)

Dimethoate Acid

**Figure 43.** Carboxylamide Hydrolysis in dimethoate.

Amidase activity is reported from various mammalian tissues, lung, muscle and pancreas but not at all from brain, spleen, and blood. It is reported that hydrolyzing activity is highest in rabbit and sheep, followed by dog, rat and steer, and low activity is found in pig, mouse, and guinea pig. Most of the hydrolytic activity was localized in both rat and sheep liver microsomes. Amidase activity was also reported from human liver and also play important role in detoxification of this compound. Enzyme activity is maximum in pH-9.0 and molecular weight is 2.3-2.5 lakh Dalton. Carboxylamidases, like the carboxylesterases, can hydrolyze only phosphorothionate insecticides and are inhibited by the corresponding phosphate analogues.

### 4.16.10. Carbamate Hydrolysis

Carbamate hydrolysis for ester group is reported by the enzyme hydrolases. It is a potential pathway for carbamate detoxification (Fig. **44**).

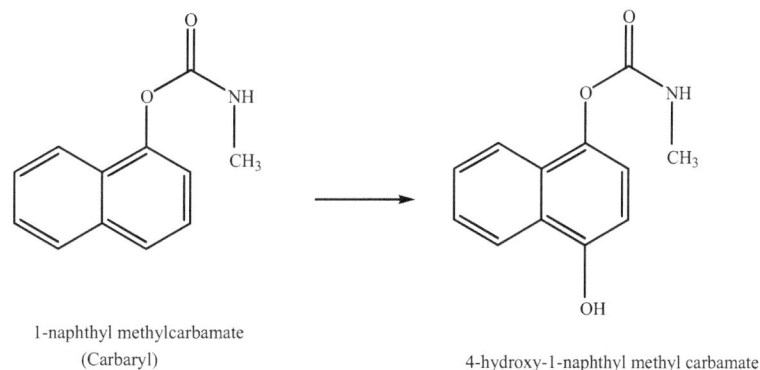

1-naphthyl methylcarbamate
(Carbaryl)

4-hydroxy-1-naphthyl methyl carbamate

**Figure 44.** Carbamate hydrolysis.

Carbaryl has been shown to hydrolyze by plasma albumin fraction from several mammalian and avian sources. The same enzyme is shown to hydrolyze the ethyl, propyl, i-propyl carbamates. It appears that in both microsomal and extramicrosomal reactions, the hydrolysis of carbamate takes place in the similar manner and it is of minor importance in carbamate metabolism.

### 4.16.11. Epoxide Hydrases

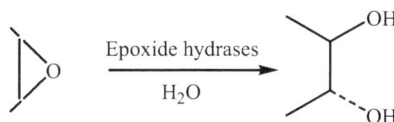

**Figure 45.** Epoxide hydrases reaction of cyclodiene pesticides.

Epoxidation is a common phenomenon of pesticide metabolism in cyclodiene insecticide where the insecticide oxidized to form epoxide rings in presence of enzyme epoxidase and they further go for hydrolysis by enzyme epoxide hydrases and form corresponding trans hydrodiols (Fig. **44**). Reactions involve here is similar as microsomal metabolism, except that enzyme is changed. This reaction is responsible for the deactivation of certain labile epoxides, which may be responsible for carcinogenesis, and also for the detoxification of certain aromatic and olefinic xenobiotics. The enzyme which mediates this reaction is highest in pig and rat and mainly associated by liver followed by kidney, intestine, and lung. Cleavage of epoxide ring of certain cyclodiene insecticide and their analogs has been well demonstrated in housefly.

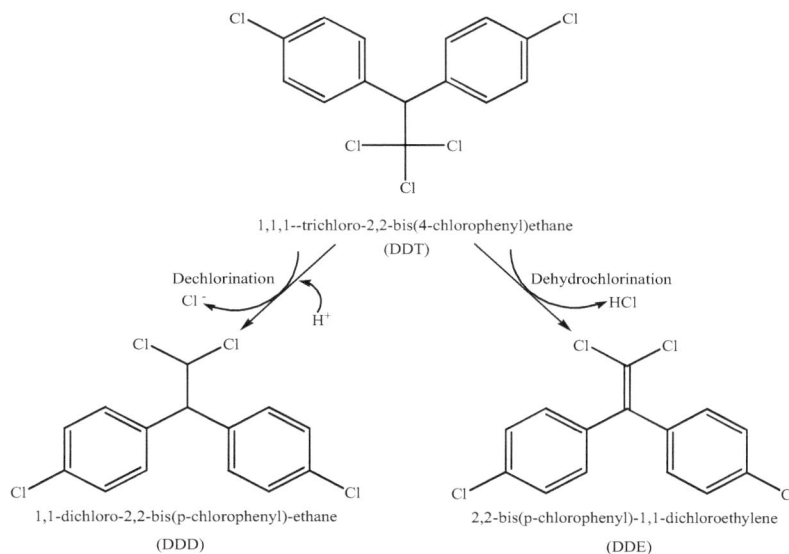

1,1,1--trichloro-2,2-bis(4-chlorophenyl)ethane
(DDT)

Dechlorination

Dehydrochlorination

1,1-dichloro-2,2-bis(p-chlorophenyl)-ethane
(DDD)

2,2-bis(p-chlorophenyl)-1,1-dichloroethylene
(DDE)

**Figure 46.** Reductive dechlorination and dehydrochlorination reactions in DDT.

## 4.16.12. Reductive Dechlorination and Dehydrochlorination

Reductive dechlorination and dehydrochlorination are the two most common metabolic pathways by which chlorinated hydrocarbon insecticide (e.g. DDT) is degraded.

Reductive dechlorination reaction is characterized by removal of chlorine atom and its replacement with a hydrogen atom e.g. dechlorination of DDT to DDD. Dehydrochlorination reaction is also characterized by the removal of chlorine atom but in this reaction, a hydrogen atom from the adjacent carbon is also removed along with the chlorine atom e.g. dehydroclorination of DDT to DDE. The enzymes responsible are dechlorinase and dehydrochlorinase which are glutathione dependent (Fig. **46**).

## 4.17. PHASE II : PESTICIDE CONJUGATION REACTIONS

The product of phase I metabolism as well as those xenobiotic compounds which contain hydroxyl, amino, carboxyl, epoxide or halogen undergoes conjugation reactions with endogenous metabolites like sugars, amino acids, glutathione, sulphate *etc.* in phase II reactions. Conjugation products are generally more polar, less toxic and more readily excretable from the body. Therefore with only few exceptions, phase II reactions are mostly detoxification reactions. Conjugation reactions can be broadly classify into three types depending upon the reactive nature of the conjugating agent and the substrate;

- **Type I conjugation reaction**: The conjugating agents are first activated by some energy compounds then combines with the substrate to form conjugates. It includes glucoside conjugation and sulphate conjugation.

- **Type II conjugation reaction**: In this case, instead of the conjugating agent, the substrates are first activated before combining with the conjugating agents to form conjugates. It includes amino acid conjugation.

- **Type III conjugation reaction**: The conjugation reaction involves no activation of either substrate or conjugating agents. The reaction thus, is proceeds directly between the substrate and the conjugating agents. It includes glutathione conjugation and phosphate conjugation.

### Learning the facts

- Phase II reaction involves conjugation of natural or foreign compounds or their metabolites with readily available, endogeneous conjugating agents *i.e.* glucuronic acid, sulfate, acetyl, methyl and glycine to form conjugates.

- Conjugation process may be viewed as a normal biochemical reaction serving as a dual role in intermediary metabolism which is responsible for detoxification of pesticides.

- Being a biosynthetic process, conjugation is generally energy dependent, so directly or indirectly linked with high energy compounds.

**Phase II reactions:**

**Type I: Pesticide / metabolite + Activated conjugating agent ⟶ Conjugated product**

- Here, pesticide conjugates with endogenous substrate which is already activated by high energy compound and finally forms a conjugated product.

- Type I reactions includes methylation, acetylation, and the formation of glucuronides, glucosides, and sulfates.

- The sites at which these enzymatic reactions occur are distributed throughout the body, although the liver constitutes the principal site.

**Type II: Activated Pesticide / metabolite + conjugating agent ⟶ Conjugated product**

- In this case, pesticide is first activated with the high energy compound and then conjugates with the conjugating agent forming the product.

- Type II reactions consist of amino acid conjugations, which occur only in the liver and or in the kidney.

**Type III: Reactive Pesticide / metabolite + reduced glutathione ⟶ Conjugated product**

- Here, the reactive pesticide or their metabolites conjugates with the conjugating agent forming conjugated product. No activation with the energy compound is required.

- In this type of conjugation, the pesticides or their metabolites possess certain chemical groups such as halogens, alkenes, $NO_2$, epoxides, aliphatic and aromatic compounds.

In general, conjugated products are ionic, polar, less lipid soluble, less toxic and easily excretable from body. Among the above three types of conjugation reactions, Type I is very common, and occurs for almost all pesticides.

### 4.17.1. Conjugation Reactions

Following are the conjugation reactions associated with pesticide metabolism.

- Glucoside conjugation

- Glutathion-s-transferase conjugation

- Sulfate conjugation

- Phosphate conjugation

- Methyl transferase conjugation

- Glycine conjugation

- Cysteine conjugation

### *4.17.1.1. Glucoside Conjugation*

Glucosidation is very important reaction in the pesticide metabolism. This type of conjugation is found both in plants and animals where reactive intermediate is derived from the universal energy fuel, glucose. The supply of this molecule is less likely to be depleted than of amino acids and other proteins.

Glucosidation is a major pathway of conjugation reaction because it has a great capacity to react with the wide range of molecules. In order to conjugate the pesticide molecule with glucose, a high energy endogenous molecules are required, which first activated the glucose and then this glucose are available for conjugation with the pesticide. This reaction is of two types,

1. *Glucoronic acid conjugation*

2. *Glucose conjugation*

## Glucoronic Acid Conjugation

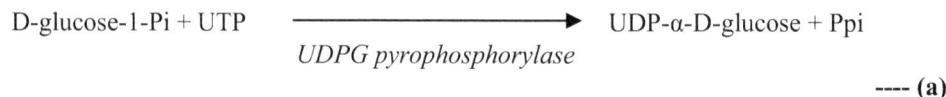

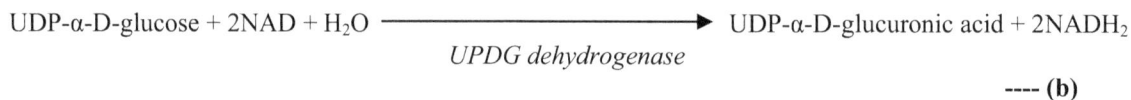

$$\text{D-glucose-1-Pi} + \text{UTP} \xrightarrow{\text{\textit{UDPG pyrophosphorylase}}} \text{UDP-}\alpha\text{-D-glucose} + \text{Ppi}$$

---- (a)

$$\text{UDP-}\alpha\text{-D-glucose} + 2\text{NAD} + \text{H}_2\text{O} \xrightarrow{\text{\textit{UPDG dehydrogenase}}} \text{UDP-}\alpha\text{-D-glucuronic acid} + 2\text{NADH}_2$$

---- (b)

In glucuronic acid conjugation, the reaction intermediate is UDPG (Uridine diphosphate Glucose), which further changes into UDPGA (Uridine diphosphate Glucuronic acid), and this conjugate with the pesticide in presence of the enzyme glucoronyltransferase.

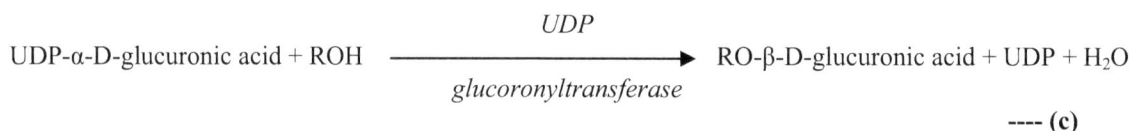

$$\text{UDP-}\alpha\text{-D-glucuronic acid} + \text{ROH} \xrightarrow[\text{\textit{glucoronyltransferase}}]{\text{\textit{UDP}}} \text{RO-}\beta\text{-D-glucuronic acid} + \text{UDP} + \text{H}_2\text{O}$$

---- (c)

Reactions (a) and (b) are catalyzed by enzymes present in the nuclear and soluble fraction of the liver, respectively. The enzyme responsible for reaction (c), UDP glucoronyltransferase is located in the microsomal fraction. Glucuronide formation occurs mainly in the liver, although other organs and tissues such as kidney, intestines, and skin also possess enzyme activity. A wide variety of chemicals can be conjugated with glucuronic acid, the most common functional groups involved being the hydroxyl, carboxyl, and amino moieties.

## Glucose Conjugation

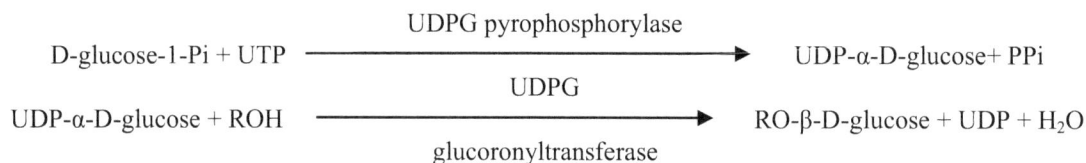

$$\text{D-glucose-1-Pi} + \text{UTP} \xrightarrow{\text{UDPG pyrophosphorylase}} \text{UDP-}\alpha\text{-D-glucose} + \text{PPi}$$

$$\text{UDP-}\alpha\text{-D-glucose} + \text{ROH} \xrightarrow[\text{glucoronyltransferase}]{\text{UDPG}} \text{RO-}\beta\text{-D-glucose} + \text{UDP} + \text{H}_2\text{O}$$

Glucose conjugation was regarded as most important reaction in pesticide detoxification, both in animals and plants. In this reaction, glucose is first activated in presence of enzyme pyrophosphorylase and then conjugated with pesticide in presence of glycoxyl transferase and form the product which is highly polar and thus excreted from body. In mammals, glycoxyl transferase is located primarily in liver microsomal fraction while in insect, it is distributed to the subcellular level.

### *4.17.1.2. Glutathion Conjugation*

This is type III reaction, here neither pesticide nor conjugating agent is getting activated, but both are reactive. The main enzyme involved in this reaction is Glutathion s transferases, which are group of enzymes that catalyze conjugation of electrophilic xenobiotic compounds with endogenous reduced glutathione. In this reaction, formation of mercapturic acid involves four important steps:

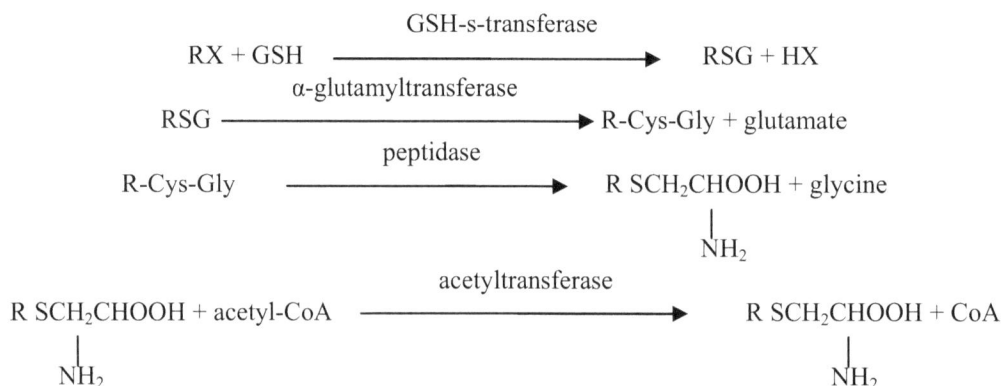

$$\text{RX} + \text{GSH} \xrightarrow{\text{GSH-s-transferase}} \text{RSG} + \text{HX}$$

$$\text{RSG} \xrightarrow{\alpha\text{-glutamyltransferase}} \text{R-Cys-Gly} + \text{glutamate}$$

$$\text{R-Cys-Gly} \xrightarrow{\text{peptidase}} \text{R SCH}_2\text{CHOOH} + \text{glycine}$$
$$\underset{\text{NH}_2}{|}$$

$$\underset{\overset{|}{\text{NH}_2}}{\text{R SCH}_2\text{CHOOH}} + \text{acetyl-CoA} \xrightarrow{\text{acetyltransferase}} \underset{\overset{|}{\text{NH}_2}}{\text{R SCH}_2\text{CHOOH}} + \text{CoA}$$

In this type of reaction, the substrate is first conjugate with reduced glutathione (GSH) in presence of enzyme Glutathion s transferase which further conjugate with cysteine and glycine to form cys-gly conjugate, and then in presence of peptidase form premercapturic acid. Subsequently cys-gly conjugate is acetylated to form mercapturic acid which becomes highly polar and eliminated in urine. This Glutathion s transferase is present in soluble fraction of mammalian liver and /or kidney, the glutamyl transferase is more active in kidney than liver. Glutathion s transferase is involved in wide variety of electrophillic insectides conjugation which can be metabolized by glutathione dependent reaction e.g. lindane, DDT and also many organophosphorus pesticide that are dealkylated or dearylated. Glutathion-s-transferase work for binding protein and serve as a storage place for toxic compound that have lipophillic nature. In certain strains of insects Glutathion-s-transferase plays an important role in development to resistance to pesticide e.g. Housefly. This reaction is also reported in mammals, reptiles, birds, fishes and invertebrates.

### 4.17.1.3. Phosphate Conjugation

Although the biosynthesis of phosphate esters is a common occurrence in intermediary metabolism, the conjugation of foreign compounds with phosphate is rarely encountered in nature. Insects appear to be the major group of animals in which phosphate conjugation has been studied to any extent. In insects phosphate conjugation is reported in several members of coleopterans, Lepidoptera and hymenoptera. It is reported that an active phosphotransferase in insects catalyzed the phosphorylation of 4-nitrophenol in the presence of ATP and $Mg^{2+}$. This enzyme is present in the high speed supernatant (100,000g) of gut tissue homogenates of the Madagascar cockroach and tobacco hornworm and of whole body homogenates of the housefly.

It is possible that ATP may serve as the activated conjugating agent in the enzymatic phosphorylation of foreign compounds by analogy with other type 1 conjugations.

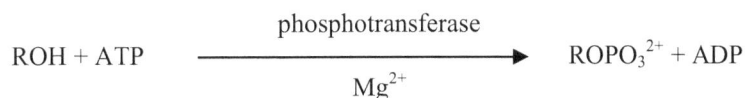

$$ROH + ATP \xrightarrow[Mg^{2+}]{phosphotransferase} ROPO_3^{2+} + ADP$$

Among mammals, phosphate conjugation is reported in human and dog. There are also reports concerning the formation of phosphate conjugates in fungi and the primitive arthropod peripatus.

### 4.17.1.4. Sulphate Conjugation

Sulphate ester formation readily occurs in phenolic hydroxyl, alcoholic hydroxyl and aromatic amino group. Sulphate ester in biological conjugation is in reality half ester which is completely ionized and highly soluble in $H_2O$. Conjugation by sulphate formation requires two stable activations. The sulphate ion is activated by ATP-sulfurylase, in the following reactions:

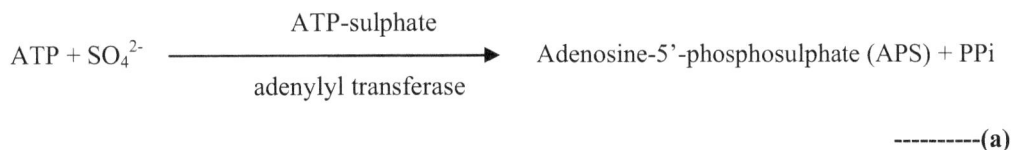

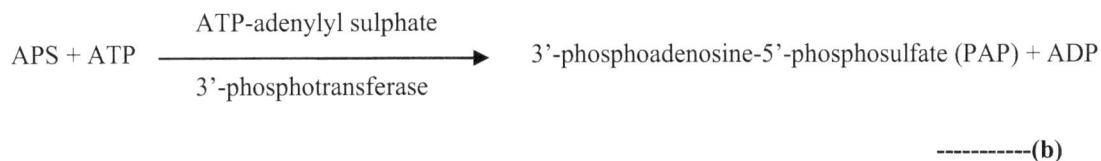

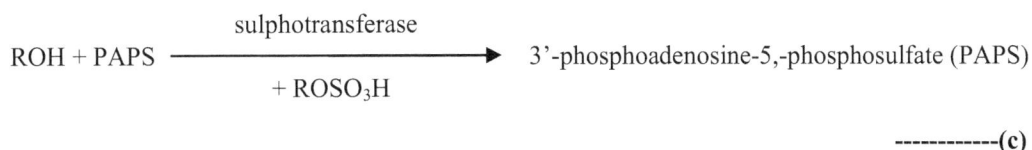

$$ATP + SO_4^{2-} \xrightarrow[adenylyl\ transferase]{ATP-sulphate} Adenosine\text{-}5'\text{-}phosphosulphate\ (APS) + PPi$$

----------(a)

$$APS + ATP \xrightarrow[3'\text{-}phosphotransferase]{ATP\text{-}adenylyl\ sulphate} 3'\text{-}phosphoadenosine\text{-}5'\text{-}phosphosulfate\ (PAP) + ADP$$

----------(b)

$$ROH + PAPS \xrightarrow[+\ ROSO_3H]{sulphotransferase} 3'\text{-}phosphoadenosine\text{-}5,\text{-}phosphosulfate\ (PAPS)$$

-----------(c)

This is the second type reaction, in which the reaction requires the biosynthesis of an active intermediate, 3'-phosphoadenosine-5,-phosphosulfate (PAPS). In the second activation reaction, APS kinase catalyzed the formation of activated sulphate PAPS through the reaction *(b)*.

Sulphate conjugates are formed by transfer of sulphate moiety from PAPS in presence of enzyme sulphotransferase in form of aryl or alkyl sulphates. Enzymes responsible for reaction (a) and (b) are located in the soluble fraction of the cell. Reaction (c) occurs with a very broad spectrum of natural and foreign substrates which include phenols, steroids, arylamines, chondroitin, choline, tyrosine methyl ester, luciferin, galactocerebroside, and heparin. It is reported that there is a family of at least 12 sulfotransferases which catalyze the PAPS-dependent sulphate conjugation in various organisms. In general, this enzyme system is located in the soluble fraction of the cell and the liver, while the presence of sulfotransferases in the gut tissues of the southern armyworm *(Prodenia eridania)* has also been reported. This enzyme system is active toward 4-nitrophenol as well as toward several naturally occurring mammalian, insect, and plant steroids, including cholesterol, $\alpha$- ecdysone, and $\beta$-sitosterol.

### 4.17.1.5. Methylation

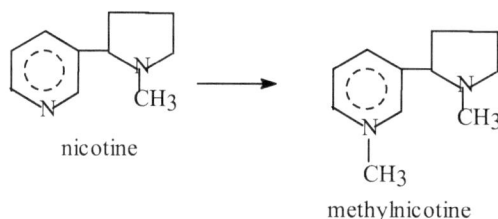

nicotine          methylnicotine

**Figure 46.** Methylation of Nicotine.

Methylation is also known as Methyl transferase reaction (Fig. **46**). Bio-methylation is an important reaction for metabolism of endogenous compounds or exogenous compounds containing O, S, and N, as a functional group. The coenzyme that is possible source of methyl group are as adenosyl methionine, 5-metyl etra hydropholic acid, vitamin $B_{12}$. The basic reaction involves the transfer of activated methyl group from S-adenosyl methionine to substrate to form methylated product and S-adenosyl homocysteine. The substrates for this reaction are primary, secondary and tertiary amines and azo hetero cycles, phenolic and thiol compounds to form nitrogen and oxygen methyl conjugates. In general, methylated products are less water soluble than their parent compound. The enzymes responsible for methylation are reported from both plants and animals. This very enzyme can further be classified as 5-adenosyl 6-methionine dependent methyl transferase or N-O adenosyl 6-methionine dependent transferase.

### 4.17.1.6. Glycine Conjugation

Aromatic and some aliphatic carbocilic acids are often conjugated with amino acid in various organisms, the most widely occurring reaction involving amino acid glycine. Glycine conjugation occurs in two stages:

1.  Activation of substrate (RCOOH) through an enzyme system which requires ATP and coenzyme A.

2.  Condensation of activated substrate with glycine.

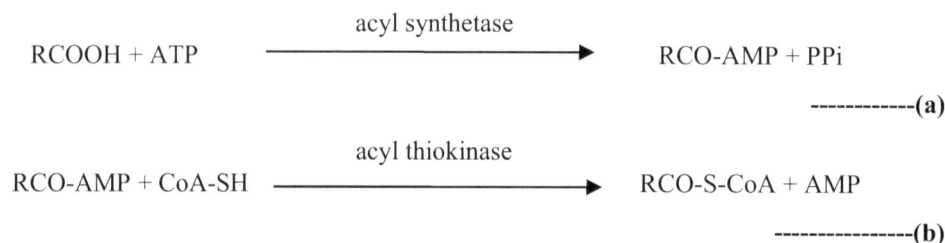

$$\text{RCO-S-CoA + Glycine} \xrightarrow[\text{Glycine N-acyl transferase}]{\text{acyl-CoA:}} \text{RCO-Gly + CoASH}$$

---------------(c)

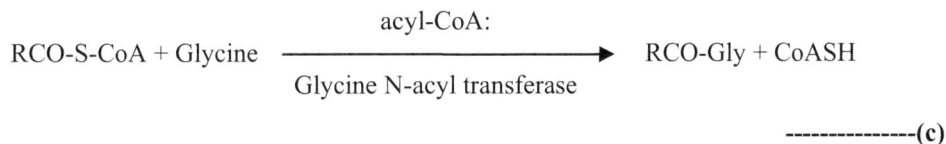

Normally these enzymatic reactions take place in the mitochondrial fraction of liver and kidney cells.

1. Rat intestinal preparations have also been shown to be active in glycine conjugation.

2. Glycine conjugation occurs in mammals, insects, amphibians, some birds, and reptiles.

3. In certain species where glycine conjugation is absent, it appears to be replaced by conjugations involving other amino acid such as ornithine, arginine, and glutamine.

### 4.17.1.7. Cysteine Conjugation or Amino Acid Conjugation

This is also a very important phase II reaction where the metabolic product of pesticide conjugate with cysteine before the formation of mercapturic acid and it become a highly polar compound and finally excreted from the body. Cysteine conjugation is mainly reported in cyclodien insecticides where they degrade by epoxidation and finally they form trans di-hydrodiol and this conjugate with amino acid cysteine at 4th position e.g. aldrin. This type of conjugation is reported in mammals, fish, bird, reptiles, plants *etc.*

### REFERENCES

[1] Williams R.T.. Detoxification Mechanism 2nd ed., Chapman and Hall, London, 1959.

[2] Ernest Hodgson and Kuhr R.J., Safer insecticides: Development and use. marcel dekker, Inc. New York, USA, (1990).

[3] Estabrook R.W., Cooper D.Y.and Rosenthal O. The light reversible carbon monoxide of the steroid C21 Hydroxylase system of the adrenal cortex. Biochem. Zeitsch. 338: 741-755, (1963).

[4] Terry Roberts and David Hutson. Metabolic pathways of agrochemicals. Part II, 1999, The Royal Society of Chemistry, UK.

### QUESTIONS

1. Describe the microsomal oxidation of the following insecticides,

   (i) Parathion

   (ii) Malathion

   (iii) Phorate

   (iv) Schradan

   (v) Chlorpyrifos

2. Describe briefly the Phase I reaction of pesticide metabolism with taking suitable examples from organophosphorus pesticides?

3. Write a short notes on,

   (i) Glucose conjugation

   (ii) Glutathion s transferase

   (iii) Methylation

*Pesticide Chemistry and Toxicology*, 2012, 97-103

# CHAPTER 5

## Toxicological Symptoms

**Abstract:** In this chapter, we have described the path of exposure to toxicants, various toxicological symptoms and their antidotes. Persons who are frequently involved with pesticides application should become familiar with the signs and symptoms of pesticide poisoning and get immediate help from a local hospital, physician, or the nearest poison control center. Treatment for poisoning with OP and carbamate insecticides involves the use of atropine which counteracts the muscarinic effects, keeping the individual alive. Antidotes are remedy or other agent used to neutralize or counteract the effects of a poison. Medical antidotes are available to neutralize the poisoning effects of the pesticides. However, if it is taken improperly then these antidotes can be more dangerous than the effects of the pesticide itself. Both carbamate and organophosphate pesticides attack cholinesterase in the blood and make it ineffective. Physician can determine the patient's base level of cholinesterase by a simple blood test. If the cholinesterase level has decreased, the patient has been overexposed to either organophosphate or carbamate pesticide. One should avoid further contact with these pesticides until his cholinesterase level has returned to normal. In severe cases, medical antidotes must be given to the patient.

Pesticides are a diverse group of substances with a potential for varied toxic effects. They can enter the human body in three ways *i.e.* by absorption, through the skin or eyes (dermally), through the mouth (orally), and by breathing into the lungs (inhalation).

**Keywords:** Exposure; Acute and Chronic toxicity; Therapy and Antidotes.

### 5.1. DERMAL EXPOSURE

Absorption of pesticide which immediately comes in contact with the skin or eyes. Absorption continues as long as the pesticide remains in contact with the skin. The *rate* at which dermal absorption occurs is different for each part of the body. The relative absorption rates are determined by comparing each respective absorption rate with the forearm absorption rate.

### 5.2. ORAL EXPOSURE

It may result in serious illness, severe injury, or even death, if a pesticide is swallowed. Pesticides can be ingested by accident, through carelessness, or intentionally.

### 5.3. RESPIRATORY EXPOSURE

It is predominantly hazardous because pesticide molecules can be rapidly absorbed by the lungs into the bloodstream. Pesticides can cause serious damage to nose, throat, and lung tissue if inhaled in sufficient amounts. Vapors and very small amount of pesticide pose the most serious risks. Lungs can be exposed to pesticides by inhalation of powders, airborne droplets or vapours. Handling concentrated wettable powders can pose a hazard if inhaled during mixing. The hazard from inhaling pesticide spray droplets is fairly low when dilute sprays are applied with low pressure application equipment. This is because most droplets are too large to remain airborne long enough to be inhaled. However, when high pressure, ultra low volume (ULV), or fogging equipment is used, the potential for respiratory exposure is increased. The droplets produced during these operations are in the mist or fog size range and can be carried along with the air currents for a considerable distance.

### 5.4. TOXICITY OF PESTICIDES: IN AN ORGANISM TOXICITY CAN BE CLASSIFIED AS

#### 5.4.1. Acute Toxicity

It refers to the effects from a single exposure or repeated exposure over a short time, such as an accident during mixing or applying of pesticides. Various signs and symptoms are associated with acute poisonings. A pesticide with a high acute toxicity can be deadly even if a small amount is absorbed. It can be measured as acute oral toxicity, acute dermal toxicity or acute inhalation toxicity.

Dileep K. Singh
All rights reserved - © 2012 Bentham Science Publishers

### 5.4.2. Chronic Toxicity

It refers to the effects of long-term or repeated lower level exposures to a toxic substance. The effects of chronic exposure do not appear immediately after first exposure and may take years to produce signs and symptoms. Examples of chronic poisoning effects may include:

1. Carcinogenicity:           Produce cancer.

2. Mutagenicity:           Cause genetic changes.

3. Teratogenicity:           Cause birth defects.

4. Oncogenicity :           Ability to induce tumor growth (not necessarily cancers).

5. Liver damage:           Death of liver cells, jaundice (yellowing of the skin), fibrosis and cirrhosis.

6. Reproductive disorders:    Reduced sperm count, sterility and miscarriage.

7. Nerve damage:           Accumulative effects on cholinesterase depression associated with organophosphate insecticides.

8. Allergenic sensitization:    Development of allergies to pesticides or chemicals used in formulation of pesticides.

The effects of chronic toxicity, as well as acute toxicity, all are dose-related. In other words, low level exposure to chemicals that have potential to cause long term effects may not cause immediate injury, but repeated exposures through careless handling or misuse can greatly increase the risk of chronic adverse effects.

Poisoning signs can be seen as vomiting, sweating or pin-point pupils. Symptoms are any functional changes in normal condition which can be described by the victim of poisoning, and may include nausea, headache, weakness, dizziness and others.

### 5.5. POISONING BY PESTICIDES

#### 5.5.1. Organochlorine pesticides

Organochlorine pesticides are not readily biodegradable and persist in the environment. These compounds may affect the nervous system as stimulants or convulsants. Nausea and vomiting commonly occur soon after ingesting organochlorine compounds.

Early signs and symptoms includes apprehension, excitability, dizziness, headache, disorientation, weakness, a tingling or pricking sensation on the skin and muscle twitching. This is followed by loss of coordination, convulsions similar to epileptic seizures and unconsciousness. When chemicals are absorbed through the skin, apprehension, twitching, tremors, confusion and convulsions may be the first symptoms.

No specific antidotes are available for organochlorine poisoning. Remove contaminated clothing immediately and then bathe and shampoo the person vigorously with soap and water to remove pesticide from the skin and hair. Persons assisting a victim should wear chemical resistant gloves and be careful to avoid becoming contaminated by the pesticide. If the pesticide has been ingested, empty the stomach as soon as possible by giving the conscious patient ipecac and water or by inserting a finger into the throat.

#### 5.5.2. Organophosphorus Pesticides

These chemical groups affect humans by inhibiting acetyl cholinesterase, an enzyme is required for proper functioning of the nervous system. The effects of OP pesticides are rapid. Symptoms commence shortly after exposure. Exposure to this class of insecticide may pose serious risks for persons with reduced lung function, convulsive disorders, *etc.* In some cases, alcoholic beverage consumption may exacerbate the pesticide effects.

Signs and symptoms associated with **mild exposures** to organophosphate insecticides includes,

1.  Headache, fatigue, dizziness, loss of appetite with nausea, stomach cramps and diarrhea,

2.  Blurred vision associated with excessive tearing,

3.  Contracted pupils of the eye,

4.  Excessive sweating and salivation,

5.  Slowed heartbeat, often fewer than 50 per minute,

6.  Rippling of surface muscles just under the skin.

These symptoms may be mistaken for those of flu, heat stroke or heat exhaustion, or upset stomach. **Moderately exposure** of organophosphate and carbamate insecticide poisoning cases exhibit all the signs and symptoms found in mild poisonings, but in addition, the victim,

1.  Is unable to walk,

2.  Often complains of chest discomfort and tightness,

3.  Exhibits marked constriction of the pupils (pinpoint pupils),

4.  Exhibits muscle twitching,

5.  Has involuntary urination and bowel movement.

**Severe poisonings** are indicated by incontinence, unconsciousness and seizures.

The order in which these symptoms appear may vary, depending on how contact is made with the pesticide. If the product is swallowed, stomach and other abdominal manifestations commonly appear first, if it is absorbed through the skin, gastric and respiratory symptoms tend to appear at the same time.

Fortunately, good antidotes are available for victims of organophosphate or carbamate poisoning at emergency treatment centers, hospitals, and many physicians' offices. As with all pesticide poisonings, time management is extremely critical. If a pesticide is swallowed, obtain prompt medical treatment. If a dermal exposure has occurred, remove contaminated clothing, wash exposed skin and seek medical care.

### 5.5.3. Carbamate Pesticides

These pesticides also are cholinesterase inhibitors (nerve poisons) and range in toxicity from low to mild toxicity. Symptoms may include,

**Mild exposure** - constricted pupils, salivation (slobbering), profuse sweating

**Moderate exposure** - fatigue, uncoordinated muscles, nausea, vomiting

**Severe exposure** - diarrhea, stomach pain, tightness in the chest.

### 5.5.4. Plant derived and Synthetic pesticides

### 5.5.4.1. Synthetic Pyrethroids

Pyrethroids are synthetically produced compounds that mimic the structure of naturally occurring pyrethrins. Some may be toxic by the oral route, but usually ingestion of pyrethroid insecticide presents relatively little risk. Very large doses may rarely cause incoordination, tremors, salivation, vomiting, diarrhea, and irritability to sound and touch. Most pyrethroid metabolites are promptly excreted by the kidney. Pyrethroids are not cholinesterase inhibitors. Crude pyrethrum is a dermal and respiratory allergen. Skin irritation and asthma have occurred following exposures. The refined pyrethrins are less allergenic, but appear to retain some irritant and/or sensitizing properties.

In cases of human exposure to commercial products, the possible role of other toxicants in the products should be considered. The synergists, such as piperonyl butoxide, have low toxic potential in humans, but organophosphates or carbamates included in the product may have significant toxicity. Pyrethrins themselves do not inhibit the cholinesterase enzyme.

Systemic toxicity by inhalation and dermal absorption is low. There have been very few systemic poisonings of humans by pyrethroids. Dermal contact may result in skin irritation such as stinging, burning, itching, and tingling progressing to numbness.

### 5.5.4.2. Rotenone

This naturally occurring substance is present in many plants. It is formulated as dusts, powders, and sprays for use in gardens and on food crops. Although rotenone is toxic to the nervous systems of insects, fish, and birds, commercial rotenone products have presented little hazard to humans.

### 5.5.5. Bioinsecticides pesticides

### 5.5.5.1. Bacillus thuringiensis (Bt)

Deliberate ingestion by human, possibility is that the organism can cause inflammation of the digestive tract. No irritation or sensitization effects have been reported in workers preparing and applying commercial products.

### 5.6. TREATMENT AND THERAPY TO PESTICIDE POISONING

Persons who are frequently involved with pesticides works should become familiar with these important steps,

1.  Recognize the signs and symptoms of pesticide poisoning.

2.  Get immediate help from a local hospital, physician, or the nearest poison control centre.

3.  Identify the pesticide to which the victim was exposed. Provide this information to medical authorities.

4.  Have a copy of the pesticide label present when medical attention is started. The label provides information that will be useful in assisting a pesticide poisoning victim.

Treatment for poisoning with OP and carbamate insecticides involves the use of atropine which counteracts the muscarinic effects, keeping the individual alive. For OP poisoning events, pralidoxime is given to prevent "aging" of the enzyme/pesticide complex. If aging is prevented, normal enzyme activity will slowly be restored as the insecticide molecule disassociates from the enzyme.

### 5.7. THERAPY AND ANTIDOTES

Antidotes are remedy or other agent used to neutralize or counteract the effects of a poison. Medical antidotes are available to neutralize the poisoning effects of the pesticides. Taken improperly, however, these antidotes can be more dangerous than the effects of the pesticide itself. Medical antidotes should be prescribed or given only by a physician. There are no known antidotes for some pesticides. Once a lethal dose has been ingested, the effects are irreversible and terminal.

The enzyme cholinesterase regulates the chemical transmission of nerve impulses, and the poison victim will die without it. Both carbamate and organophosphate pesticides attack this enzyme in the blood and make it useless. After a physician has determined the patient's base level of cholinesterase, a simple blood test will show if this level has decreased. If the cholinesterase level has decreased, the patient has been overexposed to either organophosphate or carbamate pesticide. One should avoid further contact with these pesticides until his cholinesterase level has returned to normal. In severe cases, medical antidotes must be given.

### 5.7.1. Types of Antidotes

#### 5.7.1.1. Cholinesterase Reactivators

These medications are used as antidotes to reverse the inhibition of acetylcholinesterase. The effectiveness of oxime compounds is attributed to their 2-formyl-1-methylpyridinium ions. Pralidoxime (Protopam, 2-PAM) - Nucleophilic agent that reactivates phosphorylated AChE by binding to the OP molecule. These antidotes are used to treat muscle weakness and respiratory muscle weakness in known or suspected OP exposure and must be administered into the patient within 24 h. The earlier this medication is administered, the better the result. Because it does not significantly relieve respiratory center depression or decrease muscarinic effects of AChE poisoning, concomitantly administer atropine to block effects of OP poison on these areas. Signs of atropinization might occur earlier with addition of 2-PAM to treatment regimen. Action of barbiturates potentiated by AChE inhibitors, antagonism with neostigmine, pyridostigmine, and edrophonium, morphine, theophylline, aminophylline, succinylcholine, reserpine, and phenothiazines can worsen condition of patients poisoned by OP insecticides or nerve agents. Rapid injection can cause tachycardia, laryngospasm, muscle rigidity, pain at injection site, blurred vision, diplopia, impaired accommodation, dizziness, drowsiness, nausea, tachycardia, hypertension, and hyperventilation, can precipitate myasthenia crisis in patients with myasthenia gravis and muscle rigidity in healthy volunteers, decrease in renal function increases serum drug levels because 2-PAM is excreted in urine, can transiently increase creatinine phosphokinase level, SGOT and/or SGPT levels increase in 1 of 6 patients.

#### 5.7.1.2. Anticholinergics

Such as atropine, cause pharmacologic antagonism of excess anticholinesterase activity at muscarinic receptors. Oximes reverse the inhibition of AChE and nicotinic effects, including muscle paralysis.

#### 5.7.1.3. Atropine (Atropair)

It is used for GI or pulmonary distress in known or suspected OP or carbamate poisonings. Continue until bronchoconstriction and bronchorrhea is controlled. High doses may be required in first 24 h of treatment. Treatment may be required for 48 h in severe cases.

#### 5.7.1.4. Pralidoxime (2-PAM, Protopam)

It indications include muscle weakness (especially respiratory) in known or suspected OP poisoning. Must be used early in poisoning, before OP-AChE bond has aged, to be effective. May help prevent intermediate and delayed neuromuscular and neuropsychiatric OP syndromes.

#### 5.7.1.5. Diazepam (Valium)

It depresses all levels of CNS (e.g., limbic and reticular formation), possibly by increasing GABA activity. Effects potentiated by phenothiazines, narcotics, barbiturates, MAOIs, and other antidepressants. Caution with other CNS depressants, low albumin levels, or hepatic disease (may increase toxicity), monitor for respiratory depression with high or repeated doses.

### *5.7.1.6. Lorazepam (Ativan)*

DOC for treatment of status epilepticus because persists in CNS longer than diazepam. Rate of injection not to exceed 2 mg/min. May be administered IM if IV access not available. Monitor for respiratory depression with high or repeated doses, contains benzyl alcohol, which may be toxic to infants in high doses, caution in renal or hepatic impairment, myasthenia gravis, organic brain syndrome, Parkinson disease, or inhibited benzodiazepine metabolism and clearance (e.g., in use of nicotine or cimetidine).

## 5.8. ADMINISTRATION OF THE ANTIDOTES FOR DIFFERENT GROUPS OF PESTICIDES

### 5.8.1. Organophosphates

1. Inject atropine sulphate intravenously in a dose of 2-4 mg. /kg body weight for an adult (0.04 to 0.08 mg/kg body weight for children) every 5-10 minutes until signs of atropinisation occur e.g. Dry mouth and usually dilated pupils. Maintain atropinisation for atleast 24-48 hours and carefully observe the patient as further atropinisation is stopped. It may be necessary to recommence treatment if signs of poisoning return.

2. Convulsions and anxiety can be treated with 5-10 mg. of diazepam injected intramuscularly.

3. While keeping the patient fully atropinised, administer also an oxime, if available, cholinesterase reactivator e.g. 2-PAM 1000-2000 mg/kg dody weight IM (Intramuscular) or IV (IntraVenous) for adults (25 mg/kg body weight for children), or Toxogonin (Merck) 250 mg for adults (4-8 mg/kg body weight for children). Repeat if necessary after 1-2 hours.

4. Morphine, phenothiazines, Succinylchloride, xanthenederivative, epinephrine and barbiturates are contraindicated.

### 5.8.2. Carbamates

1. Atropine therapy as indicated for organophosphorous compounds.

2. Oximes such as 2-PAM, P2S, Toxogonin should not be administered.

3. Morphine, Phenothiazines, succinylchloride, xanthenederivtive, epinephrine and barbiturates are also contraindicated.

4. Convulsions can be treated with diazepam (valium, Roche)

### 5.8.3. Organochlorines

1. Obtain and secure an unobstructed air way by suction, if necessary, from pharynx and trachea. If necessary, give artificial respiration.

2. Control convulsions by administering anticonvulsants like diazepam or paraldehyde, soluble barbiturates (Phenobarbital upto 0.7 gm per day or pentobarbital 0.25 to 0.5 gm per day)

3. 10% Calcium gluconate to be given IntraVenously.

4. In severe cases, it is necessary to protect vital organs like liver by injecting corticosteroids and kidney by dialysis.

5. Give fat free diet with high proteins, carbohydrates and calcium.

6. Adrenalin derivatives are contraindicated, since they may induce ventricular fibrillation.

7. Avoid giving morphine, theophyline or aminophylline.

8.  Patients who have had one or more convulsions should be kept under close observation for atleast 24 to 48 hours.

### 5.8.4. Synthetic Pyrethroids

1.  In case of severe skin exposure in handling or application, typical sensations of exposed skin, especially of face may appear which can be described as tingling, burning or numbness. These sensations will wear off in the course of a few hours.

2.  The treatment is symptomatic.

3.  In case larger amounts have been ingested, perform gastric lavage.

4.  Administration of activated charcoal followed by saline Cathartic with Sodium sulphate solution.

5.  Control seizures with injectable diazepam or barbiturates.

### REFERENCES

[1]  Pesticide Biochemistry and Physiology. C.F.Wilkinson (Ed.) Plenum Press, NY, 1976.
[2]  Dileep K. Singh, Toxicology of insecticides. http://nsdl.niscair.res.in/ , CSIR, New Delhi, India, 2007.
[3]  IUPAC, Compendium of Chemical Terminology (The Gold Book), Compiled by A.D.McNaught and A. Wilkinson, Blackwell Science, Oxford, 2nd edn, 1997.
[4]  Fundamental Toxicology, ed. J.H.Duffus and H, G.J. Worth, Royal Society of Chemistry, Cammbrifge, 2nd edn. 2006.

### QUESTIONS

1.  Write a short notes on

    (a)  Acute toxicity

    (b)  Chronic toxicity

    (c)  Mutagenicity

    (d)  Carcinogenicity

    (e)  Antidotes for organophosphorus pesticides

    (f)  Cholinesterase reactivators

2.  Describe briefly the therapy which can be used at the time of organophosphorus pesticide poisoning?

# CHAPTER 6

# Pesticide Formulation and Application

**Abstract:** In this chapter different pesticide formulations and its method of application has been described. Pesticides are utilized in the form of formulations. Formulation is final physical condition in which the insecticide is sold commercially. Formulation is the processing of a pesticide by such methods that will improve its properties of storage, handling, application, effectiveness and safety to the applicator and the environment, and profitability. Many formulations are made like emulsifible concentrate, water-miscible liquid, wettable powder, water soluble powders, oil solutions, flowable powders, aerosols, granular, fumigants, ultra low volume concentrates, fogging concentrates, dusts, impregnated materials, poison baits, slow release insecticides *etc.* A technical grade of pesticide is the pure form (purity up to 95-98 per cent) of the chemical.

Pesticides are applied in the field with the help of suitable equipment. A variety of sprayers are available in the market. Granule applicators as well as dusters are used for soil treatment. Ultra low volume application became possible through the development of equipments that allow application of very narrow range of droplet sizes. The application patterns, day timing and temperature are important considerations while applying the pesticides. The type and level of pest infestation decide type of the pesticides to be use as well as the pesticide application method.

**Keywords:** EC; Granules; Dust; SL; WML; WP; WSP; FP; Aerosol; Fumigants; ULV; Poison Baits; Slow release formulation; Sprayers; applicators; Insecticide Act;

## INTRODUCTION

Proper application of pesticide in the field is necessary for maximizing the pesticide performance. Various pesticide formulations like Emulsifiable Concentrate (EC), Solubilised Liquid (SL), Granules (G), Dusts, Wettable Powders (WP), Oil Solutions, Aerosol and Ultra Low Volume Concentrates are available in the market. They are easy to handle and are safe to use. Most insecticides are sold to the users as emulsifiable concentrates having a shelf life of approximately 3 years.

Pesticide application equipments come in all shapes, sizes, types and prices. Various types of sprayers and nozzles are available. Granule applicators as well as dusters are also used for pesticide application. These application equipments provide a uniform coverage of the field. Functioning and handling of equipments are usually known to the farmers/applicators. Temperature, timing and the day length are also important factors at the time of pesticide application. Here we will discuss the pesticide application in special reference to India.

India is basically an agricultural country with most variable climatic conditions and different geographic features. A variety of cereals, oil seeds, pulses, vegetables and horticultural crops are being cultivated in the country. Therefore, India requires a major share of different pesticides to be used in order to increase its productivity.

Import, manufacture, sale, transport and usage of pesticides are being regulated under a comprehensive statute " The Insecticides Act, 1968" and the rules framed there are under to ensure availability of quality, safe and efficacious pesticides to the farming community. The Agricultural Universities, Plant Protection Division, Central Insecticide Board, Registration Committee and Central Insecticide Laboratory work under this act and perform their duties. Various proposals are processed depending on the funds made available for these purposes by the Government of India.

On the contrary, extensive use of pesticides is detrimental to the environment. The ultimate fate of the pesticide being introduced through an application in the environment is affected by many processes. These processes determine insecticide persistence as well as its movement. Non-judicious use of pesticide can cause plant injury or excessive residues which contribute to the environmental contamination. These processes include transfer (movement) and breakdown of pesticide. Pesticide has many impacts on air, water, soil and non-target organisms.

Dileep K. Singh
All rights reserved - © 2012 Bentham Science Publishers

Also, due to the non-judicious use of insecticide the insect pests have started developing resistance to the insecticide applied over a period of time against them. Development of resistance means a major economic loss as well as increased hazards due to higher quantities of pesticides used. So, to prevent insecticide resistance as well as to protect the environment a number of strategic management practices under the "Integrated Pest Management Practices" are recommended.

## 6.1. INSECTICIDE FORMULATION

After an insecticide is manufactured in a relatively pure form (technical grade), it must be formulated before it can be applied. Formulation is the processing of the technical grade by various methods which is done to make the product safer, more effective and more convenient to use. In a formulation, there are one or more chemicals (formulants) which are the active ingredients (a.i.) and other ingredients which have no pesticide action (inert ingredients). Inert ingredients may be toxic to the applicator irrespective of the fact that they have no pesticide action.

There are mainly three types of pesticide formulations (liquid, solid and gas). A single pesticide may be sold in more than one formulation. Some products are ready to use and require no further mixing. However, most products applied in the liquid form must be diluted in water or oil before use. Formulation type depends on several factors,

- toxicology of the active ingredient,

- chemistry of the active ingredient,

- how effective the product is against the pest,

- the effect of the product on the environment (plant, animal or surface *etc.*),

- how the product will be applied and the equipment needed the application rate

### 6.1.1. Characteristics of an Appropriate Insecticides Formulation

- **Highly toxic to target insects:** Generally, with due course of time, target insects develop resistance to the insecticide and thus insecticide may lose their effectiveness. If resistance is observed another insecticide without cross-resistance has to be used.

- **Not repellent or irritant to target insects:** To ensure that the insects pick up a lethal dose.

- **Long-lasting:** The toxicity should remain high over a sufficiently long period to prevent the need for frequent reapplication, which is costly and time-consuming.

- **Safe to humans and domestic animals:** There should be no danger to spray workers, inhabitants or animals accidentally contaminated with the insecticide during or after spraying.

- **Stable during storage and transportation:** Insecticides should be stable during storage and transportation.

- **Cost-effective:** Insecticide should be economical in terms of the money spent on it.

There are various types of insecticide formulations which are commonly used are:

### 6.1.2. Emulsifiable Concentrate (EC)

Emulsifiable concentrates are concentrated liquid formulations in which the water insoluble insecticide is dissolved in a suitable solvent. Emulsifier is a surface-active agent which is partly hydrophilic and partly lipophilic and is an important component of these formulations. More than 75% of all insecticide formulations are applied as sprays. When an emulsifiable concentrate is added to water, the emulsifier causes the oil to disperse uniformly throughout the water phase, giving an opaque or milky appearance when agitated. This oil-in-water suspension is a normal emulsion. A few formulations are invert emulsions, that is, water-in-oil suspensions. These are opaque and thick, concentrated emulsions resembling salad dressing or thick cream, and are employed almost exclusively as herbicide formulations. Constant agitation during spraying is required to keep the spray solution uniformly mixed.

### 6.1.3. Granules

The size of granules ranges between 0.1 and 2.5 mm and they are solid particles. It includes the technical product combined with an inert carrier. The granules are safe, easy to handle product which allows for precise targeting and slow release of the active ingredient. Granular formulations are primarily used soil treatments and may be applied either directly to the soil or over plants. When applied over plants, the granules fall through the foliage to the soil. Some granules are applied to water either directly or over foliage. Since the granules do not adhere to the foliage of most plants, phytotoxicity and residue problems are reduced.

### 6.1.4. Dusts

The insecticide powders are made by adding a solution of the active ingredient in a suitable solvent to a finally ground carrier. The particles must be of the optimum size; if too small, they will cling together in the air blast and the coverage of the target surface will be poor. On the other hand, if too coarse the dust will not penetrate into the interior of plants and the active ingredient may separate from the particles of diluents, this does not occur if the diameters of the particles are all less than 20 μm. Improperly blended or mixed dusts will not spread the toxicant evenly on the treated surface. If the dust is unusually gritty, the material was not ground sufficiently. This can also cause uneven distribution of the formulated material to plants or other treated surfaces.

### 6.1.5. Solubilised Liquid

Solubilisation refers to the mixing of a water-soluble pesticide with oil by using suitable surfactants as co solvents. With this method, a formulation is produced which increases penetration through the bark or leaf cuticle, but permits translocation of the active ingredient in the plant. The technique has been tried with some herbicides.

### 6.1.6. Water Miscible Liquids

The miscibility of technical grade material is in water or alcohol. Their resemblance is with emulsifiable concentrates in viscosity and colour, but they do not form milky suspensions when diluted with water. They are harmful and very few insecticides are formulated in this manner.

### 6.1.7. Wettable Powders (WP)

These are dry or powdered preparations containing surface-active or wetting agents and usually other conditioning materials. When a wettable powder is mixed with water or other liquids, a suspension spray is formed. Coarseness or grittiness will be abrasive and cause damage to pumps and nozzles of the equipment.

### 6.1.8. Water-Soluble Powders

These powders contain a finely ground solid which dissolves readily upon the addition of water. They may contain a small amount of wetting agent to assist their solution in water. Unlike wettable powders, they do not require constant agitation and form no precipitate. Very few insecticides are of this kind.

### 6.1.9. OIL SOLUTIONS

In these formulations, the insecticide is dissolved in an organic solvent and then used in insect control. It is rare that they are used on crops because they can cause severe burning of foliage. They are used effectively on livestock, as weed sprays along road sides, in standing pools for mosquito control. The concentrated solutions are diluted with kerosene or diesel oil before application.

### 6.1.10. Flowable Powders

Few insecticides are soluble neither in oil nor in water. In this case, the technical product is wet-milled with a clay diluents, water, a suspending agent, a thickener and an anti-freeze compound thus forming a thick creamy mixture which mixes well with water. But it needs frequent agitation while application.

### 6.1.11. Aerosols

Aerosols are particles dispersed in gas. The liquefied-gas aerosols consist of a toxicant dissolved in a liquid and held under pressure in a container. When the valve is opened, a fine spray emerges. The liquid solvent quickly vaporizes leaving the small toxic particles suspended in the air. In general, there are space aerosols for controlling flying insects and contact aerosols for control of ants and roaches. The main difference is the amount and type of toxicant dispersed. The particles in the contact aerosol are large and heavy, therefore would not remain suspended in the air.

### 6.1.12. Fumigants

Fumigants may be volatile liquids, solids or gases or mixtures of them which produce gas, vapour, fumes or smoke intended to destroy pests such as insects, rodents, nematodes, weed seed, plant pathogens, *etc.*

### 6.1.13. Ultra-Low Volume Concentrates (ULV)

It usually contains a technical product dissolved in a minimum amount of solvent. Their application requires no further dilution. It is applied in volumes of 0.61 to a maximum of 4.7 l/ha in very small sprays droplets of 1-15 μ. The advantage of ULV sprays is that the small droplets can better penetrate thick vegetation and other barriers. These formulations are used where insect control is desired over large areas.

### 6.1.14. Impregnated Materials

These chemicals are used in treatment of woolens for moth-proofing and timber against wood- destroying organisms.

### 6.1.15. Poison Baits

Formulated baits contain low levels of toxicants incorporated into materials such as foodstuffs, sugar, molasses, *etc.* that are attractive to the target pest. These are purchased commercially or they can be formulated by pest control operators.

### 6.1.16. Slow Release Insecticides

These formulations are relatively new and only a few are available commercially. An example is the Shell-No-Pest Strip, which incorporates dichlorvos (DDVP) into strips of polychlorovinyl resin. The insecticide volatilizes slowly, killing flying and crawling insects over a long period of time.

### 6.2. METHODS AND EQUIPMENTS FOR PESTICIDE APPLICATION

The application patterns, volume applied and droplet size distribution are definitely influenced by the type of equipment used for applying the insecticide. So, an outline of this subject is required. The application of insecticides is still dominated by water based sprays. These sprayers forcing the liquid insecticide with pressure through the nozzle produce a sheet of insecticide that breaks into droplets of various sizes and volumes. Ultra low volume application became possible through the development of equipments that allow application of very narrow range of droplet sizes.

The decision to adopt the insecticide application method of insect pest control will not only depend on the type and level of pest infestation but also very importantly on the farmers. Every user will opt that insect control method which is easy to use, safe and economically affordable to him.

## 6.2.1. Types of Sprayers

### 6.2.1.1. Proportioner or Hose-End Sprayer

They are inexpensive small sprayers which are designed to be attached to a garden hose. A small amount of insecticide is mixed with water and placed in the receptacle attached to the hose. A tube connects this concentrate to the opening of the hose. When the water is turned on, the suction created by the water passing over the top of the tube pulls the insecticide concentrate up and into the stream of hose water. The stream can reach into medium-high trees if water pressure is high.

### 6.2.1.2. Trombone Sprayer

The trombone sprayer is a medium-sized, hand-held piece of equipment. A spray mixture in the correct dilution is prepared in a container such as a bucket. The intake tube of the sprayer is inserted into the mixture in the bucket. Pump pressure is created by operating the sprayer in a trombone-like motion. The insecticide is pulled up the hose and out *via* the end of the sprayer. When wettable powder is used, the spray mixture is frequently agitated to keep it in suspension. Trombone sprayers are excellent for spraying trees and shrubs.

### 6.2.1.3. Compressed Air Sprayer (backpack or tank sprayer)

Spray is mixed in a small tank (generally 1 to 5 gallons) and the tank is carried over the shoulders. A hand operated pump supplies pressure during application. Frequent agitation of the spray mixture is necessary when using a wettable powder formulation. Applicator has excellent control over coverage, making this sprayer a good choice for treating dwarf fruit trees, vegetables, and ornamentals. But the spray will not reach to tall trees.

### 6.2.1.4. Small Power Sprayers

They are motor-driven sprayers and are lightweight. Power sprayers provide uniform pressure, but are generally expensive.

### 6.2.1.5. Hydraulic Nozzles

Most of the world's insecticides are mixed with water and sprayed through <u>hydraulic nozzles</u> of one sort or another. This is more than 100 years old technology and is still considered the method of choice, by most farmers and other spray operators.

### 6.2.1. 6. Rotary Nozzles

Rotary nozzles are normally used to achieve Controlled Droplet Application (CDA). CDA is the term used to describe a new method of applying pesticides. Controlled droplet application (CDA) technology produces spray droplets that are relatively uniform in size and permits the applicator to control droplet size and it is a reliable way of applying insecticides at ultra-low volume (ULV) rates of application.

### 6.2.1.7. Granule Application

Granule application can be as simple as distributing the formulation to plant bases or central whorls with a (gloved) hand. Granular formulations are frequently prepared for toxic insecticides but manual application equipment is not recommended for poisonous (especially class I) compounds.

The use of the some instrument as illustrated, can aid spreading, control flow and reduce operator contamination.

### 6.2.1.8. Application of Dusts

Dusts are extremely small particles and are liable to be inhaled. The dusting of crops is becoming less frequent, but application equipment, hand duster is still available.

### 6.2.1.9. Hand Duster

The duster may consist of a squeeze tube or shaker, a plunger that slides through a tube, or a fan powered by a hand crank. Uniform coverage of foliage is difficult to achieve with many dusters. Dusts are more subjected to drift than liquid formulations due to their light weight and poor sticking qualities.

## 6.3. PROPER PESTICIDE APPLICATION STRATEGIES

Application equipment should be checked for leaking hoses or connections before application. Insecticides should be applied only on days with no breezes to minimize drift. Early morning is considered as the safest time of day to spray to reduce the hazard of drift. High temperatures increase vaporization of the insecticides which can cause harm to the environment. So, spraying should be done during the cool part of the day to reduce vaporization.

## 6.4. INSECT PESTS AND RECOMMENDED INSECTICIDES

### 6.4.1. Application of Different Pesticides Commonly used in India

| S.No. | Pesticide Name | Used Against |
|---|---|---|
| 1. | DDT | Effective against wide variety of insects, including domestic insects and mosquitoes |
| 2. | Endosulfan | It is used as a broad spectrum non systemic, contact and stomach insecticide, and acaricide against insect pests on various crops |
| 3. | Aldrin | Effective against wireworms and to control termites |
| 4. | Dieldrin | Used against ectoparasites such as blowflies, ticks, lice and widely employed in cattle and sheep dips. Also used to protect fabrics from moths, beetles and against carrot and cabbage root flies/ Also used as seed dressing against wheat and bulbfly |
| 5. | Heptachlor | It controls soil inhibiting pests. |
| 6. | Chlordane | It is a contact, stomach and respiratory poison suitable for the control of soil pests, white grubs and termites. |
| 7. | Lindane | It is used against sucking and biting pest and as smoke for control of pests in grain sores. It is used as dust to control various soil pests such as flea beetles and mushroom flies. It is effective as soil dressing against the attack of soil insects |
| 8. | Fenitrothion | It is a broad spectrum contact insecticide effective for the control of chewing and sucking pests- locusts, aphids, caterpillars and leaf hoppers. It is also used against domestic insects and mosquitoes |
| 9. | Fenthion | It is a persistent contact insecticide valuable against fruitflies, leaf hoppers, cereal bugs, and weaverbirds in the tropics |
| 10. | Parathion | A contact insecticide and acaricide with some fumigant action. Very effective against soil insects with high mammalian toxicity |
| 11. | Profenofos | Used for control of important cotton and vegetable pests. Used against chewing and sucking insects and mites, cotton bollworms, aphids, cabbage looper and thrips |
| 12. | Phorate | A systemic and contact insecticide employed for the control of aphids, carrot fly, fruit fly and wireworm in potatoes |
| 13. | Malathion | Widely used insecticide and acaricide used for the control of aphids thrips, red spider mites, leafhoppers and thrips |
| 14. | Monocrotofos | A powerful contact and systemic insecticide and acaricide with a broad spectrum of activity used to control pests on crops like cotton, rice, soyabean, maize, coffee, citrus and potatoes |
| 15. | Dimethoate. | A systemic and contact insecticide and acaricide, effective against red spider mites and thrips on most agricultural and horticultural crops |
| 16. | Chlorpyrifos | A broad spectrum insecticide used against mosquitoes, fly larvae, cabbage root fly, aphids, codling and wintermoths on fruit trees. It is also used in homes, restaurants against cockroaches and other domestic pests. It is also used for the control of termites |

| 17. | Diazinon | A contact insecticide effective against a number of soil, fruit, vegetable and rice pests e.g. cabbage root, carrot and mushroom flies, aphids, spidermites, thrips and scale insects domestic pests and livestock pests |
| 18. | Quinalphos | A broad spectrum contact and systemic insecticides applied as a spray to control pests in cereals, brassicas and other vegetables |
| 19. | Triazophos | Used against flies and insect pests of cerealos, maize, oilseed rape, brassicas, carrots, weevils in peas and cut worms in potatoes and other crops |
| 20. | Ethion | Used for the control of aphids and mites |
| 21. | Acephate | It is a systemic insecticide effective against chewing and sucking pests. |
| 22. | Fenvalerate | It act contact and stomach poison. It controls the pests on crops of cotton, vegetables and fruits. |
| 23. | Permethrin | It is a stomach and contact insecticide effective against broad range of pests of cotton, fruit and vegetable crops. |
| 24. | Cypermethrin | It is a stomach and contact insecticide effective against broad range of pests of cotton, fruit and vegetable crops. |
| 25. | Deltamethrin | It is a potent insecticide effective as a contact and stomach poison against broad range of pests of cotton, fruit and vegetable crops and store products. |
| 26. | Carbaryl | It is a contact insecticide and fruit thinner with a broad spectrum of activity effective against many pests of fruits, vegetables and cotton. It is also used to control earthworms and leather jackets in turf. |
| 27. | Carbofuran | It is a broad spectrum systemic insecticide, acaricide and nematicide used against insects, mites and incorporated in soil for control of soil insects and nematodes. |
| 28. | Aldicab | It is a systemic insecticide, acaricide and nematicide which are formulated as granules for soil incorporation. It is effective for control of aphids, nematodes, flea beetles, leaf miners, thrips and white flies on a wide range of crops. |
| 29. | Methomyl | It is used as a soil and seed systemic insecticide applied as a foliar spray to control aphids. |
| 30. | 2, 4-D | It is a selective systemic post emergence herbicide used for the control of many annual broadleaf weeds in cereals, sugarcane and plantation crops. |
| 31. | Butachlor | It controls annual grasses and some broad leaved weeds in transplant and direct seeded rice. It is applied as pre-emergence in EC formulations and as early post emergence in the form of granules. |
| 32. | Paraquat | It is used as a plant desiccant effective against grasses. |
| 33. | Simazine &Atrazine | It is a persistent soil acting herbicide which in high concentrations acts as total weed killer and in lower concentrations is used for selective control of germinating weeds in a variety of crops - maize, sugarcane, pineapple, sorghum. It is also used for long term control of annual grass and broad-leaved weeds in crops like citrus, coffee, tea and cocoa. |
| 34. | Glyphosate | It is a potent non-selective post emergence herbicide which kills mono and dicotyledonous annual and perennial weeds |
| 35. | Isoproturon | It is used to control annual grass weeds in wheat rye and barley. |
| 36. | Trifluralin | It is used for the control of annual grasses and broad leaved weeds in a wide range of crops cotton, groundnuts, soyabeans, brassica, beans and cereals. |
| 37. | Mancozeb | It is a protective fungicide, effective against a wide range of foliage disease. |
| 38. | Captan | It is a foliage fungicide with protective action. It is mainly used for seed treatment and soil fungicide. |
| 39. | Captafol | It is a protective, wide spectrum foliage and soil fungicide. |
| 40. | Carbendazim | It is a systemic fungicide which controls wide range of pathogens of cereals, fruits, grapes ornamentals and vegetables. It is very effective against leaf and ear disease of wheat. |

Source: Pollution Monitoring Laboratory, CSE

The discovery of DDT in 1939 brought a revolution in pest control. Thereafter, the chemical control became popular due to immediate knock-down effects. Today, the use of chemical pesticide remains one of the best known and most widely used pest control tactics. India is predominantly a pesticide market. Introduction of high yielding varieties together with increasing application of agrochemicals increased the productivity of land with a concomitant increase in the proportion lost to insect pests in India and other developing Asian countries. In advanced countries, organochlorine group of pesticides has been banned but India still uses some of these products. But after the ban on the use of DDT and HCH, the situation has changed and the share of insecticides has declined. In 1995, 67% of DDT and BHC were used but the percentage declined to 51.6% in 1997-98.

Conventional pesticides were successful in controlling insect pests during the past five decades, minimizing thereby losses in agricultural yields. Unfortunately many of these chemicals are harmful to man and beneficial organisms and cause ecological disturbances. In India at present 186 insecticides have been registered under the Section 9 (3) of Insecticide Act, 1968 (March, 2005). Total of 12 biopesticides are registered. Pesticide banned for manufacture, import and use are 24 in number and 4 pesticide formulations are banned for import, manufacture and use. Total of 3 pesticide/ pesticide formulations are there which are banned for use but still, their manufacture is allowed for export and total of 7 pesticides are restricted. Besides, adverse effects on environment, the drawbacks in use of the pesticides include development of resistance and increasing costs of pesticides. Overall excess use of pesticides led to resurgence of minor pest, ecological imbalance, residue pesticide in food and environmental pollution.

In 1962, Rachel Carson published her book "Silent Spring" where she has given emphasis on misuse and overuse of pesticides on the environment. So, the concept of Integrated Pest management (IPM) emerged. Integrated Pest management (IPM) is defined as a pest management system that utilizes suitable techniques and methods against the pests in as compatible manner with the environment as possible and thus, maintaining the pest population levels below those causing economic injury. So, IPM is always in the context of the environment and population dynamics of the pest species. Thus IPM is the best combination of all possible approaches in pest management with least reliance on chemical pesticide.

Thus ultimate objectives of IPM are,

1.   Reduce management cost.

2.   Minimize environmental pollution.

3.   Maintain ecological balance with minimum disturbance to ecosystem.

## 6.5.  INSECTICIDES RECOMMENDATIONS, APPLICATIONS, REGULATIONS AND CONTROL AGENCIES

The use of synthetic pesticides has increased steadily because of the spreading use of new high-yielding varieties of seeds which are very vulnerable to plant pests and diseases. The sale of synthetic pesticides jumped from 8,620 tons in crop year 1960-61 to 85,660 tons in 1989-90. In 1960 only about 6 million hectares received chemical pesticides, but by the early 1980s some 100 million hectares were being treated, and the growth in coverage continues in the 1990s.

But the pesticides are poison causing both acute and chronic health effects. Therefore, on the one hand the use of pesticides increases the production of food but on the other side it also creates a lot of health problems. Annually thousands of people are dying due to pesticide poisoning. So, the rapid rise in the use of plant protection led to the enactment of the Insecticides Act of 1968 to regulate the import, sale, transport, distribution, and use of insecticides.

The ministry of agriculture monitors the manufacture, sale, import, export and use of pesticides through the "Insecticide Act, 1963". Ministry of Agriculture regulates pesticides usage in the country. Because of the death of over 100 people due to pesticide poisoning in Kerala in 1958 the Insecticide Act, was enacted. Acts and rules are primarily focused towards regulating the acute health effects of pesticides.

For the formation of new insecticide the companies firstly approach the Agriculture University. These Universities test the new insecticide by studying several aspects of insecticides like the efficacy, toxicity and MRLs [Maximum Residue Limits] *etc.* In the test regarding the efficacy it is measured that with what capacity an insecticide can kill the insect and how effective it is for application. Along with it the toxicity of insecticide to human being is also analyzed. For MRL fixing of the Insecticide the data collected by the Agriculture University is sent to Department of Health under PFA [Prevention of food Adulteration Act, 1954].

Performing all these norms the Agriculture University let this information to the concerned company to approach the Central Insecticide Board and Registration Committee. The Registration Committee checks all these information in their meeting and recommends the name of the insecticide to the Agriculture Ministry, Govt. of India. Then the Central Government as per the powers conferred by sub clause [ii] of clause [e] of section 3 of the Insecticide Act, 1968 [46 of 1968], publish the name of insecticide in the Gazette of India and after that the Registration Committee issue the license to the concerned company.

The Central Insecticides Board (CIB), constituted under Section 4 of the Act, renders advice to the central and state governments on the technical matters arising out of the administration of the act and to carryout such other functions. The Registration Committee (RC) constituted under Section 5 of the Act registers insecticide under section 9 of the Act, after satisfying itself about their efficacy and safety to human beings, animals and environment.

The Central Insecticide Laboratory (CIL), Faridabad set up under Section 5 of the Act, is serving as a referral laboratory for quality control of pesticides. The CIL is also carrying out pesticides residues analysis and investigations on bioassay, medical toxicology and processing /packaging. Besides there are two Regional Pesticides Testing Laboratories (RPTLs) at Kanpur and Chandigarh to assist the states in quality control tests.

A central task force has been created in the Department of Agriculture and Cooperation to organize and coordinate raids across the country to ensure supply of quality pesticides to farmers. The said Task Force initiated such raids in Delhi, Haryana, Punjab and Uttar Pradesh, whereby samples of pesticides were drawn from various dealers and suppliers for the purpose of testing.

## 6.6. PESTICIDE CONTROL LEGISLATION IN INDIA

| No. | Legislation |
|---|---|
| 1 | Insecticide Act, 1968 and the rules framed under it Insecticide Rules, 1971 |
| 2 | Environment Protection Act, 1986 |
| 3 | Prevention of food Adulteration Act, 1954 |
| 4 | Factories Act, 1948 |

According to the above mentioned laws the agricultural ministry and Central Insecticide Board control the use of insecticides in our country. Several pesticides like Aldrin, Aldicarb, Benzene Hexachloride (BHC), Calcium Cyanide, Chlordane Copper Acetoarsenite, Dibromochloropropane (DBCP), Endrin, Ethyl Mercury Chloride, Ethyl Parathion, Heptachlor, Menazon, Nicotine Sulphate*, Nitrofen, Paraquate Dimethyl Sulphate, Pentachloro Nitrobenzene (PCNB), Phenyl Mercury Acetate (PMA)*, Tetradifon and Toxaphene *etc.* are banned for use in India.

In India, at present 186 insecticides have been registered and 7 pesticides are restricted. Pesticide banned for manufacture, import and use are 24 in number and 4 pesticide formulations are banned for import,

---

*These are manufactured in India for export purposes only.

manufacture and use. Total of 3 pesticide/ pesticide formulations are there which are banned for use but still, their manufacture is allowed for export.

Judicious use of quality pesticides is an indispensable input in plant protection. Pesticides are commonly used in plant protection measures for sustaining food production. These are also used for the control of vector born diseases. Pesticides, being toxic by their very nature, are hazardous to human beings and ecosystem. The residues of pesticides enter into food chain and cause harm to human and animal health.

## REFERENCES

[1]     Dileep K. Singh, Toxicology of insecticides. http://nsdl.niscair.res.in/ , CSIR, New Delhi, India, 2007.
[2]     Fred Whitford. The Complete Book of Pesticide Management. John Wiely & Sons Ltd. 2002.

## QUESTIONS

1.   What are the insecticide control regulations in India? How will you compare it from European countries?

1.   What you understand about restricted use and banned pesticides?

2.   Describe briefly the usage of different sprayers in pesticide application and their benefits?

**CHAPTER 7**

# Pesticides and Environment

**Abstract:** In this chapter, the behavior of pesticides in soil, water and air environment has been described. When the pesticide is introduced in the environment it transferred to different components of the ecosystem and move in different trophic levels. Transfer includes adsorption, leaching, volatilization, spray drift, and runoff. Pesticide also degrades in environment by microbes (microbial breakdown), chemical reactions (chemical breakdown), or light (photo degradation).

Extensive use of insecticides may lead to the development of resistance in insect. Resistance is defined as the ability to develop tolerance to doses of an insecticide which would prove lethal to the majority of individuals in a normal population of the same species. It is a dynamic phenomenon that has multiple factors which are broadly classified as biochemical, physiological, morphological, genetic and ecological factors.

Development of resistance means a major economic loss as well as increased hazards due to higher quantities of pesticides used. So, to prevent pesticide resistance as well as to protect the environment a number of strategic management practices are recommended.

**Keywords:** Absorption; Leaching; Volatilization; Mineralization; Half life; Degradation; Hazards; Resistance.

## INTRODUCTION

The insecticides introduced in the environment through an application or a disposal is affected by many processes. These processes determine insecticide persistence, its movement and its ultimate fate. The fate processes can be beneficial or detrimental. Beneficial, when the insecticide moves to the target area and detrimental when there is reduced control of a target pest, injury of non-target plants and animals and environmental damage.

These processes include transfer (or movement) and degradation of the insecticides. Transfer includes processes that move the insecticide away from the target site. These include adsorption, leaching, volatilization, spray drift, and runoff. Insecticide has many impacts on air, water, soil and non-target organisms (Fig. **1**).

That is why, some insecticides have been banned due to the fact that they are persistent toxins which have adverse effects on environment. A classic example which is often quoted is that of DDT. DDT has been widely used (and maybe misused) insecticide. DDT is responsible for the thinness of the egg shells of predatory birds. The shells sometimes become too thin to be viable, causing reductions in bird populations. This occurs with DDT and a number of related compounds due to the process of bioaccumulation.

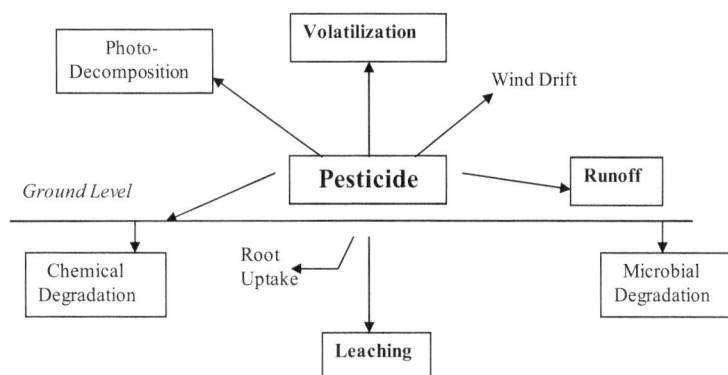

**Fig. 1.** Behaviour of pesticide in environment

Bioaccumulation is defined as the phenomenon in which the chemical, due to its stability and fat solubility, accumulates in fat of the organisms. A similar term is biomagnification which is defined as a phenomenon

**Dileep K. Singh**
**All rights reserved - © 2012 Bentham Science Publishers**

in which the insecticides get stored in body fats that can be ingested by animals (including humans) higher up the food chain. Insecticide persistence is measured in terms of the half-life, or the time in days required for an insecticide to degrade in soil to one-half of its original amount. For example, if an insecticide has a half-life of 16 days, 50 percent of the insecticide applied will still be present 16 days after application. The longer the half life, the persistent the insecticide.

## 7.1. FATE OF PESTICIDES IN THE ENVIRONMENT

### 7.1.1. Adsorption

The adsorption process is a phenomenon which binds insecticides to soil particles because of the attraction between a chemical and soil particles. Positively charged insecticide molecules, for example, are attracted to and can bind to negatively charged clay particles.

Moisture affects adsorption. Dry soils absorb more insecticide than wet soils because water molecules compete with the insecticide for the binding sites in the wet soils. Many soil factors influence the insecticide adsorption. Soils high in organic matter or clay are more adsorptive than coarse, sandy soils, in part because a clay or organic soil has more particle surface area, or more sites onto which insecticides can bind.

Due to persistency of insecticides, some insecticides stay in the soil long enough to be absorbed by plants grown in the field later. They may damage or leave residues in future crops.

### 7.1.2. Leaching

Leaching is a phenomenon of the movement of insecticides through the soil rather than over the surface. An insecticide that dissolves in water can move with water in the soil. So, solubility is a factor. Soil permeability (how readily water moves through the soil) is also important. The more permeable a soil (for example, sandy soil), the greater is the potential for insecticide leaching.

An insecticide held strongly to soil particles by adsorption is less likely to leach. The likelihood of leaching also depends on how much persistent the insecticide is. The insecticide with a low persistence (that is rapidly broken down by a degradation process) is less likely to leach because it may remain in the soil only for a short time. Typically, the closer the time of application to a sustained or heavy rainfall, the greater the chances that some insecticide will leach.

### 7.1.3. Volatilization

Volatilization is the conversion of a solid or liquid into a gas. Once volatilized, an insecticide can move in air currents away from the treated surface. Vapour pressure is an important factor in determining whether an insecticide will volatilize or not. The higher the vapour pressure, the more volatile the insecticide will be.

Some other environmental factors tend to increase volatilization. They include high temperature, low relative humidity, and air movement. An insecticide tightly adsorbed to soil particles is less likely to volatilize; soil conditions such as texture, organic matter content, and moisture can thus influence insecticide volatilization.

### 7.1.4. Spray Drift

Spray Drift is the airborne movement of spray droplets away from a treatment site during application.

### 7.1.5. Runoff

Runoff is the movement of insecticides in water over a sloping surface. Insecticides may move as compounds dissolved in the water or attached to soil particles of the eroding soil. It depends on the slope or grade of an area; the erodibility, texture and moisture content of the soil. The amount and timing of rainfall and irrigation are also important factors.

Runoff can also occur when water is added to a field faster than it can be absorbed into the soil. Over-irrigation can lead to accumulation of excess surface water which causes insecticide runoff.

Runoff from areas treated with insecticides can pollute streams, ponds, lakes, and wells. Insecticide residues in surface water can harm plants and animals and contaminate groundwater.

---

**Learning the Facts :** Half life of any pesticide can be determined by an equation

$$T\frac{1}{2} = \frac{Ln2}{Ln(Po/P)} \times t$$

T½        = Half Life

Ln2       = 0.693

Ln        = Natural Log

Po        = Initial Concentration

P = Final Concentration

T = Time (period)

---

## 7.2. DEGRADATION

Degradation is the process of breaking down of pesticides by microbes, chemical reactions, or light. This process may take from hours or days to years, depending on environmental conditions and the chemical characteristics of the insecticide. It provides the concept of half life of the pesticides in the environment.

### 7.2.1. Microbial Degradation

It is the degradation of pesticide by microorganisms such as fungi and bacteria. When conditions like, adequate oxygen and soil moisture, warm temperature and favourable pH are present then microbial breakdown tends to increase.

### 7.2.2. Chemical Degradation

It is the degradation of pesticides by chemical reactions in the soil. The rate and type of chemical reactions that occur are influenced by soil temperature, pH levels, moisture and the binding of insecticides to the soil.

### 7.2.3. Photodegradation

It is the degradation of pesticides by sunlight. All insecticides are susceptible to photodegradation to some extent. The rate of breakdown is depending on the intensity of light, length of exposure, and the properties of the insecticide.

## 7.3. PESTICIDE MOBILITY IN THE ENVIRONMENT

Insecticides that move away from the release site may cause environmental contamination which is very harmful. Insecticides move in several ways, including:

- in air, through wind currents.

- in water, through runoff or leaching.

- on or in plants, animals, humans *etc.*

### 7.3.1. Air

Insecticides drift away from the release site in the air. Also, fumigant insecticides are intended to form a vapour when they are released. Vaporization is the evaporation of an active ingredient during or after application. Insecticide vapours move about easily in the air. These form a source of air pollutants which are harmful as either directly or in a changed form during subsequent chemical reactions.

### 7.3.2. Water

#### 7.3.2.1. Ground Water

It was believed for years that the natural filtering of water during its slow movement through the soil, sand, gravel and rock was sufficient to cleanse it of contaminants before it reached groundwater. Today, many chemicals, including some insecticides, have been detected in groundwater. Groundwater is contaminated if insecticides leach from treated fields, mixing sites, washing sites, or waste disposal areas.

#### 7.3.2.2. Surface Water

The surface water system such as rivers, lakes, streams, reservoirs and estuaries are especially vulnerable to the accumulation of insecticides and other chemicals because they are small captive sinks of the by-products of man's activities. Surface water is linked to both groundwater and atmospheric water through the hydrologic cycle. Surface water moves into groundwater by percolating downwards through the soil and it also enters the atmosphere through evaporation and transpiration. Likewise, water from the atmosphere and groundwater can recharge surface waters. The insecticides reach surface water systems by direct application, spray drift, atmospheric fallout (rain and dust storms) and run-off from agricultural land.

### 7.3.3. Soil

The capacity of the soil to filter, degrade and detoxify insecticides is a function or quality of the soil. The presence of insecticides in soil can adversely impact soil organisms, beneficial plants, animal as well as human health. Insecticides can move off-site contaminating surface and groundwater and possibly causing adverse impacts on aquatic ecosystems.

### 7.3.4. Non-Target Organisms

Non-target organisms constitute all the organisms excluding the target insect pest species against which the insecticide has been applied. It includes wildlife, birds, aquatic ecosystem, honeybees, beneficial insects as well as natural enemies of the insect pests. Insecticides are harmful to the non-target organisms in two ways:

- the insecticides may harm by contacting the non-target organisms directly, or

- the insecticides may leave a residue which may harm later.

### 7.3.5. Harmful Effects from Direct Contact

Organisms can be exposed to insecticides directly by eating contaminated food or water, breathing insecticides, or by skin absorption.

The insecticide toxicity and insecticide quantity (dose) are the two factors on which the type and intensity of the effects depend upon. If exposure causes the organism's death, it is referred to as a lethal effect.

### 7.3.6. Harmful Effects from Residues (Indirect Effects)

A residue is the part of an insecticide remaining in the environment after an application. The more persistent the insecticide, more the residue it will leave which is harmful.

Organisms may also experience lethal or sublethal effects without being directly exposed to an insecticide. This typically occurs when an insecticide application destroys the food sources or disrupts the food chain.

### 7.3.7. Hazards to Wildlife

In general the risk an insecticide poses to wildlife is related to the insecticide type, its toxicity, the dose, number of applications, and the nearness of the application site to wildlife habitat, the insecticide persistence, and its ability to concentrate in the wildlife food chain (biomagnification). These factors interact with food habits and behavior of individual wildlife species to produce a response.

### 7.3.8. Hazards to Bird Population

The decline in the number of rapacious birds such as peregrine falcon has been greatest in the areas where persistent organochlorine pesticides (DDT and dieldrin) have been widely applied. These birds of prey occupy the top most position in the food chain. And due to the phenomenon of biomagnification, they accumulate a higher dose of the organochlorines. Recent findings have shown that the persistent pesticides induce liver enzymes that lower the estrogen levels in the birds.

Also, residues of organochlorine insecticides and their metabolites in bird eggs have been implicated in the thinning of egg shells with a consequence of egg breakage or reproductive failure. DDE, a metabolite of DDT is the main factor in eggshell thinning. The inhibition of carbonic anhydrase by DDT and its metabolites appears to play a major role in the thinning of eggshells and egg breakage.

### 7.3.9. Hazards to Aquatic Ecosystem

Inland bodies of water such as groundwater, rivers, lakes, streams, reservoirs and estuaries are especially vulnerable to the accumulation of insecticides and other chemicals. The insecticides reach aquatic systems by direct application, spray drift, atmospheric fallout (rain and dust storms) and run-off from agricultural land. Carried away by natural forces such as wind, rain, dust storms, the flow of rivers and ocean currents, residues of insecticides began to appear everywhere on the globe from tropical forests to Antarctic snows.

The progressive use of insecticides has reached to massive quantities in the past 40-50 years and has caused detrimental effects on aquatic forms of life. The effects of insecticides pollution on aquatic systems depend on the chemical characteristics of the compound, its stability and persistence, its water solubility, its potential for uptake and bioconcentration into aquatic organisms.

The most stable insecticides, such as the organochlorine compounds, slow and decomposing, has taken heavy tolls of many fish species, aquatic invertebrates and a wide variety of non-target organisms. The gradual build-up of organochlorine residues in fish is a long term threat, with the consequential hazards to humans who stand last at the food chain.

### 7.3.10. Hazards to Honeybees

Honeybees are very important because they are the major pollinators. Some insecticides may kill honeybees, causing severe economic losses to beekeepers and loss of certain crops due to poor pollination. The danger of insecticides to bees results not only from direct contact with poison, but also from taking poisoned nectar, pollen and water, and their transport to the hive.

Honeybees are more severely damaged if the harmful insecticides are applied when bees are foraging in the field for nectar and pollen, so application of materials toxic to bees should be avoided when crops are in bloom. Usually, insecticidal dusts are more hazardous than sprays, and oil solutions and concentrates are more hazardous than water emulsions and suspensions.

### 7.3.11. Hazards to Beneficial Insects

Most insecticides will have an effect on some beneficial insects. Beneficial insects include predators and parasites of insect pests, pollinators, and the food sources for other useful species. The mortality of beneficial insects is an important consideration when spraying is done in complex ecosystems such as forests or agricultural crops that are established over a period of years (e.g. alfalfa). Any insect in the upper crop canopy at the time of spraying will be exposed to a potentially lethal dose of insecticide at the time of application. Insects entering fields which have been treated with insecticide shortly can be harmed.

## 7.3.12. Hazard to Natural Enemies of Insect Pests

The concept of integrated pest management emphasizes the maximum use of natural enemies of pests supplemented with selective insecticides when necessary. The insecticides, both persistent and nonpersistent types, have detrimental effects on the balance between pest populations and their natural enemies.

The most common example is that of Indonesia. The extensive increase in the use of insecticide from 1980-1985 resulted in destruction of natural enemies of the major insect pest: brown plant hopper. This led to severe failure of the rice production changing Indonesia from exporter to importer. Then 57 out of 64 insecticides used on rice were withdrawn and this caused rise in rice yields as in the absence of pesticidal effect the population of the natural enemies started increasing.

## 7.4. INSECT RESISTANCE TO PESTICIDES

Resistance is defined as the ability to develop tolerance to doses of an insecticide which would prove lethal to the majority of individuals in a normal population of the same species. Resistance is a genetically based decrease in susceptibility of a population to an insecticide. It is an evolutionary phenomenon because resistance requires changes in allele frequencies.

Insecticide resistance in insect is considered to be an important problem in increasing the productivity. Resistance develops mainly because of the extensive use of insecticides to control insect pests.

### 7.4.1. History of Resistance

In, 1911, there was evidence that extensive fumigation of citrus trees with hydrogen cyanide (HCN) to control citrus scales (an insect pest) had selected a resistant strain of insects to this compound.

### 7.4.2. Development of Resistance

The development of resistance by insects is due to the selection of variants in the population carrying preadaptive genes for resistance. Insect populations usually become resistant to an insecticide used to control them only after the majority of susceptible insects have been removed by that particular insecticide. The remaining insects that carry genes for resistance, mate with each other and the resulting resistant insect populations subsequently increase in numbers, making their control difficult and this further leads to the development of resistance. The rate at which resistance develops in a population depends on:

1.   The frequency of resistant genes present in the population

2.   Nature of this genes (either single or multiple, dominant or recessive)

3.   Intensity of selection pressure

4.   The reproductive potential of the population *i.e.*, number of generations per year

Initially, when insecticide is applied to control the insect pests then satisfactory control is achieved. At this time the frequency of resistant insects is very low (one in million or more). With the repeated use of the same insecticide, the frequency of resistant insects increases leading to occasional crop losses. But the continuous use of the insecticide leads to the tremendous increase in the growth rate of the population of the resistant insect pest and the insecticide becomes ineffective. This is called as insecticide resistance.

### 7.4.3. Factors of Insect Resistance

It is a dynamic phenomenon that has multiple factors which are broadly classified as biochemical, physiological, morphological, genetical and ecological and ethological factors. Factors involved in resistance are:

### 7.4.4. Biochemical and Physiological Factors

It is the most important factor in developing resistance. The basic mechanism behind the developing resistance among the insects could be due to:

1.  An increased detoxification of the insecticides by the breakdown of the chemical by enzymes such as:

    a.  The cytochrome mediated P-450 mediated mixed-function oxidases which introduces oxygen to the molecule.

    b.  Esterases which hydrolyses esters and amides making the compound more polar

    c.  Glutathione S-Transferases which catalyse the conjugation of the molecule with the thiol (SH) group of glutathione.

2.  Breakdown products of parent compound into less toxic or nontoxic components which can be excreted easily.

3.  Altered target site of action of the insecticide leads to the insensitivity of the insecticide.

### 7.4.5. Morphological Factors

Such as insect cuticle reduce the penetration of the insecticide into the insect's body.

### 7.4.6. Genetic factors

Several recessive genes like **kdr gene** (knockdown resistance gene) are present in different frequencies in a population, are responsible for developing resistance among the insect population.

### 7.4.7. Ecological Factors

Geographical location is one of the considerations of the insect resistance.

### 7.4.8. Ethological factors

Sometimes behavioral resistance is also observed among the insects. For example, certain strains of mosquitoes and houseflies avoid resting on surfaces treated with DDT or Malathion.

### 7.4.9. Cross resistance

Cross-resistance occurs when selection with one insecticide causes resistance to another insecticide. In other words, cross resistance is a specific mechanism which confers resistance to two or more compounds with the involvement of the same gene. The main reason behind the cross resistance is the same mode of action or similar metabolic pathway for different compounds. For example, insects resistant to organophosphorus insecticides generally become cross resistant to carbamate insecticides.

### 7.4.10. Insects and Insecticide resistance management

Various species and genus of mosquitoes *viz., Culex* sp., *Anopheles* sp. *etc.* have become resistant to a variety of organochlorine and organophosphate insecticides like DDT, malathion, dieldrin, HCH, *etc.* Major insect pests of the agricultural crops have also developed resistance to various insecticides. For example, Diamond back moth (*Plutella xylostella*) a serious pest of crucifers throughout the world has developed resistance to nearly all classes of insecticides. Another serious pest, *Helicoverpa armigera* has shown insecticide resistance towards pyrethroids, various organochlorine, and organophosphate as well as to carbamates insecticides. Tobacco caterpillar, *Spodoptera litura,* a very common pest has also become resistant to a variety of insecticides. The real development of pesticide resistance in this pest took place during 10 years period from 1971 to 1980. Similarly, a vast majority of cases related to insecticidal resistance have been reported in various agricultural, veterinary, public health and household pests.

### *7.4.10.1. Management strategies*

Intensive use of insecticides led the insecticide resistance an inevitable consequence. Development of resistance means a major economic loss as well as increased hazards due to higher quantities of pesticides used. So, to prevent insecticide resistance as well as to protect the environment a number of strategic management practices are recommended which are classified as follows:

- Judicious use of insecticides - insecticides should be used only if the use is very essential, that is after monitoring the pest population in the field.

- Rotation or alteration of the different insecticides or the use of mixtures of insecticides – use of different insecticides with different mode of action can nullify each other's resistance.

- Use of synergist enhances the toxicity of the insecticide by inhibiting the detoxification mechanism.

**7.4.10.2. Synergy** or *synergism* (from the Greek *synergos* meaning working together) refers to the phenomenon in which two or more or agents acting together create an effect greater than the sum of the effects each is able to create independently. One or the other combination of the above prescribed methods will definitely lead to the suppression of the build-up of insect resistance.

### REFERENCES

[1]    Pesticide Biochemistry and Physiology. C.F.Wilkinson (Ed.) Plenum Press, NY, 1976.
[2]    Dileep K. Singh, Toxicology of insecticides. http://nsdl.niscair.res.in/ , CSIR, New Delhi, India, 2007.

### QUESTIONS

1.  Define the following terms,

    (i)    Synergism

    (ii)   Cross resistance

    (iii)  Volatilization

2.  Describe briefly, the hazards of organochlorine pesticide to the bird population?

3.  How will you calculate the half life of pesticide in terrestrial environment?

4.  What you understand about development of resistance for pesticide in insect? What will be your strategies to overcome it?

# CHAPTER 8

# Summary

**Abstract:** In this chapter, the main theme of pesticide usage and its regulations are summarized in general to make some understanding about the pesticide toxicology.

**Keywords:** Pesticides; Behaviour in environment.

## INTRODUCTION

India is an agrarian country providing employment to around 67% population of the country. Improved farming techniques and the use of high-yielding varieties have significantly increased the production. Also, the use of chemical fertilizers and pesticides has played a positive role in increasing agricultural productivity.

The pesticide industry in India is growing rapidly. About 3% of the total pesticides used in the world are utilized in India. The use pattern of pesticides suggests that cotton ranks first with 44.5% of the total consumption followed by rice, which accounts for 22.8% pesticide consumption. These two crops consume more than two thirds of the total quantity of pesticides in the country. While among the states Tamil Nadu, Andhra Pradesh, Uttar Pradesh and Maharashtra are using maximum quantity of pesticides. These four states alone utilize about 48.8% of the total pesticides.

Insecticides are utilized in the form of formulations. Formulation is the processing of an insecticide by such methods that will improve its properties of storage, handling, application, effectiveness and safety to the applicator and the environment and profitability. Many formulations are made like emulsifible concentrate, water-miscible liquid, wettable powder, water soluble powders, oil solutions, flowable powders, aerosols, granular, fumigants, ultra low volume concentrates, fogging concentrates, dusts, impregnated materials, poison baits, slow release insecticides *etc.* A technical grade of insecticide is the pure form (purity up to 95-98 per cent) and it is always depicted as active ingredient (a.i.) in formulations.

Insecticides are applied in the field with the help of suitable equipment. A variety of sprayers are available in the market. Granule applicators as well as dusters are also used for solid formulations. Ultra low volume application became possible through the development of equipments that allow application of very narrow range of droplet sizes. The application patterns, day timing, temperature *etc.* are important considerations while applying the insecticides. The type and level of pest infestation decide the insecticides to be use as well as the insecticide application method.

Although, use of high yielding varieties, intensive agricultural practices, high doses of fertilizers and chemical pesticides brought significant increase in the yield of crop production. But still, crop losses are the subject of speculation and research. The yield loss due to insect pests is a very important issue. So, there is an urgent need to assess such losses, in order to frame strategies to overcome them. The elaborate data of the losses due to insect pests depend on many factors and is therefore, quite difficult to evaluate correctly. These factors are the intensity of infestation, time and duration of attack, density of pest population, plant growth stage and the cultural practices followed.

In 1939, the discovery of DDT brought a revolution in pest control. Thereafter, the chemical pesticides became very popular due to immediate knock-down effects. In 1962, the Rachel Carson published her book "Silent Spring" which brought about the world wide debate over the use of pesticides and the related environmental concern. Then the concept of Integrated Pest management (IPM) emerged. Integrated Pest management (IPM) is defined as a pest management system that utilizes suitable techniques and methods against the pests in as compatible manner with the environment as possible and thus, maintaining the pest population levels below those causing economic injury. So, the ultimate objective of IPM is to maintain ecological balance with minimum disturbance to ecosystem keeping the management cost in view.

**Dileep K. Singh**
**All rights reserved - © 2012 Bentham Science Publishers**

Institutes like CIB-RC (Central Insecticide Board and Registration Committee), Agricultural Universities, Department of Health and Ministry of Agriculture are playing important role in recommendation of insecticides so that they can be applied in order to enhance the crop yield. These institutions are formed in order to regulate the use of insecticides. Few insecticides are restricted for use in India. Like use of DDT is banned in agriculture. Use of Lindane formulation generating smoke for indoor use is prohibited in India. So all this work of registration, regulation and application of insecticides is monitored by the above mentioned organizations and they work in a proper interlinked manner. Various proposals are processed depending on the funds available for these purposes by the Government of India.

When an insecticide is introduced in the environment, many processes affect its ultimate fate. The fate processes can be beneficial or detrimental. Beneficial, when the insecticide moves to the target area and detrimental when there is reduced control of a target pest, injury of non-target plants and animals and environmental damage. These processes include transfer, movement, breakdown and degradation. Transfer includes processes that move the insecticide away from the target site. These include adsorption, leaching, volatilization, spray drift, and runoff. Insecticide has many impacts on air, water, soil and non-target organisms. Degradation is the process of breaking down of insecticides by microbes (Microbial breakdown), chemical reactions (Chemical breakdown), or light (Photodegradation).

Today, many chemicals, including some insecticides, have been detected in groundwater. Groundwater may be contaminated if insecticides leach from treated fields, mixing sites, washing sites, or waste disposal areas. The presence of insecticides in soil can adversely affect the soil organisms, beneficial plants, animal as well as human health. Insecticides can move from the site contaminating surface and groundwater and possibly causing adverse impacts on aquatic ecosystems.

Insecticidal impact to non-target organisms includes direct as well as indirect effects. Organisms can be exposed to insecticides directly by eating contaminated food or water, breathing insecticides, or by skin absorption. Organisms may also experience lethal or sublethal effects without being directly exposed to an insecticide. This typically occurs when an insecticide application destroys or disrupts food sources. Egg shell thinning is caused due to the inhibition of carbonic anhydrase by DDT and its metabolites which led to significant fall of the reproduction rate in rapacious birds. Honeybees are more severely damaged if the harmful insecticides are applied when bees are foraging in the field for nectar and pollen.

So, the concept of integrated pest management emphasizes the maximum use of natural enemies of pests supplemented with selective insecticides when necessary. The insecticides, both persistent and nonpersistent types, have detrimental effects on the balance between pest populations and their natural enemies.

Extensive use of insecticides led to the development of resistance in insect. Resistance is defined as the ability to develop tolerance to doses of an insecticide which would prove lethal to the majority of individuals in a normal population of the same species. It is a dynamic phenomenon that has multiple factors which are broadly classified as biochemical, physiological, morphological, genetic and ecological factors. Development of resistance means a major economic loss as well as increased hazards due to higher quantities of pesticides used.

## REFERENCE

[1]　Dileep K. Singh. Insecticidal method of pesticide application. http://nsdl.niscair.res.in/ , CSIR, New Delhi, India, 2007.

# Terminologies

**Abiotic:** The nonliving components of a system, such as temperature, water, soil type, sunlight and air pollutants.

**Absorption:** The assimilation of molecules into cells, the process by which plants take up nutrients through their roots.

**Acaricide:** A pesticide toxic to mites and ticks.

**Acceptable daily intake (ADI):** The amount of a substance in food or drinking water, expressed on a body mass basis (usually mg/kg body weight), which can be ingested daily over a lifetime without considerable health risk.

**Accumulation:** Successive additions of a substance to a target organism, or organ, or to part of the environment, resulting in an increasing amount or concentration of the substance in the organism, organ, or environment

**Acetylcholine**: A chemical that functions as a synaptic neurotransmitter in the nervous system of animals.

**Active ingredient (a.i.):** Chemicals in a pesticide formulation that is biologically active as toxins. A given pesticide formulation may have more than one active ingredient.

**Acute effect:** Effect in a short duration and occurring rapidly (usually in the first 24 h or up to 14 d) following a single dose or short exposure to a substance or radiation.

**Additive effect:** The efficacy of a pesticide mixture that is equal to the sum of the toxicities of the individual pesticides (*i.e.* antagonism and synergism).

**Adjuvant:** Any nonpesticidal substance in a pesticide formulation that improves the physical, chemical, or biological properties of the active ingredient, or improves application efficacy.

**Administration (of a substance):** Application of a known amount of a substance to an organism in a reproducible manner and by a defined route.

**Adsorption:** The process that takes place when a liquid or a gas (adsorbate) accumulates on the surface of a solid (adsorbent), forming a molecular or atomic film. The exact nature of the bonding depends on the characteristic nature of the chemical species involved.

**Adverse effect:** Change in morphology, physiology, growth, development or lifespan of an organism which results in impairment of functional capacity or impairment of capacity to compensate for additional stress or increase in susceptibility to the harmful effects of other environmental influences.

**Aerobic:** A process or an organism that requires oxygen.

**Aerosol:** Very fine droplets (0.1 to 5 μm in diameter) suspended in air, generated by a container pressurized with a gas propellant, or aerosol generators such as fogging machines or ultra-low-volume (UVL) equipment.

**Agitate:** Stirring or shaking a pesticide mixture so that the components will not separate or settle in the application tank.

Dileep K. Singh
All rights reserved - © 2012 Bentham Science Publishers

**Antibiotic:** Chemicals produced by microbes that inhibit, or are toxic, to other microbes.

**Antibody:** Protein molecule produced by the immune system (an immunoglobulin molecule) which can bind specifically to the molecule (antigen or hapten) which induced its synthesis

**Anticoagulant:** A substance that inhibits blood clotting resulting in internal hemorrhaging, this is the mode of action of a major class of rodenticides.

**Antidote:** Substance capable of specifically counteracting or reducing the effect of a potentially toxic substance in an organism by a relatively specific chemical or pharmacological action.

**Antigen:** Substance or a structural part of a substance which causes the immune system to produce specific antibody or specific cells and which combines with specific binding sites (epitopes) on the antibody or cells

**Antifeedant:** Materials that inhibit or stop pest feeding. A control method often used for clothes moth larvae and termites.

**Asphyxia:** Condition resulting from insufficient intake of oxygen: symptoms include breathing difficulty, impairment of senses and, in extreme, convulsions, unconsciousness and death.

**Asphyxiant:** Substance that blocks the transport or use of oxygen by living organisms.

**Ataxia:** Unsteady or irregular manner of walking or movement caused by loss or failure of muscular coordination.

**Attractant:** Chemical substances or devices used to lure insects or other mobile pests to areas where they can be trapped or killed. Attractants are based on feeding, oviposition, or mating behaviour of insects.

***Bacillus thuringiensis (Bt):*** Soil-inhaling bacterium that produces an insecticide effective against larval stages of many species of Lepidoptera, although some strains are effective against various beetle, mosquito, and black fly larvae.

**Bacterial insecticide:** Bacteria pathogenic to insects (e.g., Bt or B. popilliae). Applied using application techniques also used for chemical pesticides.

**Bait:** A pesticide formulated with an attractive food substance and containing a small amount of toxic active ingredient (usually about 5%). Baits are primarily used for control of mollusks, cutworms, and rodents.

**Bioaccumulation:** Progressive increase in the amount of a substance in an organism or part of an organism which occurs because the rate of intake exceeds the organism's ability to remove the substance from the body.

**Bioactivation:** Any metabolic conversion of a xenobiotic to a more toxic derivative.

**Bioassay**: an experiment in which test organisms are exposed to different

**Bioconcentration:** Process leading to a higher concentration of a substance in an organism than in environmental media to which it is exposed.

**Biodegradation:** Decomposition or breakdown of a substance through the action of microorganisms (such as bacteria or fungi) or other natural physical processes (such as sunlight).

**Biological control or biocontrol**: Regulating pest populations by using natural enemies such as herbivores, pesdators, parasitoids, and parasites.

**Biomagnification:** Sequence of processes in an ecosystem by which higher concentrations are attained in organisms at higher trophic levels (at higher levels in the food web), at its simplest, a process leading to a higher concentration of a substance in an organism than in its food.

**Biomineralization:** Complete conversion of organic substances to inorganic derivatives by living organisms, especially micro-organisms.volume, in a medium such as water.

**Biotic insecticide**: natural or introduced enemies of a pest including predators and parasites that are applied using standard pesticide application techniques.

**Biopesticides:** Certain types of pesticides derived from organisms as animals, plants, bacteria, and certain minerals. For example, Bacillus thuringiensis, a bacterium which is currently being used as biopesticides on a large scale.

**Biotransformation:** Any chemical conversion of substances that is mediated by living organisms or enzyme preparations derived there from.

**Botanicals:** Pesticides derived from plants, such as pyrethrum, rotenone (derris) ryania, and nicotine.

**Brand name:** The name that manufacturers use for commercial purposes.

**Carcinogen: A Chemical or Biological Agent which Produces, Accelerates or Increases Frequency of Cancers.**

**Carcinogenicity:** Process of induction of malignant neoplasms by chemical, physical or biological agents

**Carrier or diluents**: An inert material used in the formulation of sprays and dust. It is usually combined with a toxicant to dilute or otherwise make the mixed material more suitable for field application.

**Chemical breakdown:** The breakdown of insecticides by chemical reactions in the soil. The rate and type of chemical reactions that occur are influenced by soil temperature, pH levels, moisture and the binding of insecticides to the soil.

**Central nervous system:** The part of the nervous system that consists of the brain and the spinal cord.

**Chemical Name: The Name that Specifies the Chemical Structure of a Compound.**

**Cholinesterase inhibitor:** Substance which inhibits the action of acetylcholinesterase and related enzymes which catalyse the hydrolysis of choline esters: such a substance causes hyperactivity in parasympathetic nerves.

**Chronic effect:** Consequence which develops slowly and has a long-lasting course (often but not always irreversible).

**Combined effect of poisons:** Simultaneous or successive effect of two or more poisons on the organism by the same route of exposure.

**Conjugate:** Derivative of a substance formed by its combination with compounds such as acetic acid, glucuronic acid, glutathione, glycine, sulfuric acid *etc.*

**Compressed Air Sprayer** (backpack or tank sprayer): The one in which spray is mixed in a small backpack tank and a hand-operated pump supplies pressure during application. A uniform concentration spray can be maintained.

**Controlled Droplet Application (CDA):** The term used to describe a new method of applying pesticides. Controlled droplet application (CDA) technology produces spray droplets that are relatively uniform in size and permits the applicator to control droplet size and it is a reliable way of applying insecticides at ultra-low volume (ULV) rates of application.

**Contaminant:** A substance that is either present in an environment where it does not belong or is present at levels that might cause harmful (adverse) health effects.

**Cross-resistance**: A specific mechanism which confers resistance to two or more compounds with the involvement of the same gene.

**Cytochrome P-450:** Haemoproteins which form the major part of the enzymes concerned with the mono-oxygenation of many endogenous and exogenous substrates. The term includes a large number of iso-enzymes which are coded for by a superfamily of genes. Endogenous substrates of these enzymes include cholesterol, steroid hormones and the eicosenoids, the exogenous substrates are xenobiotics. Strictly, the cytochrome P450 family are not cytochromes but are haem-thiolate proteins.

**Cytochrome P450 oxidase:** A generic term for a large number of related, but distinct, oxidative enzymes in animal physiology. The cytochrome P450 mixed-function oxidases system is an important element of Phase I metabolism in animals which are responsible for the chemical modification or degradation of chemicals including pesticides, drugs and endogenous compounds.

**Cytoplasm:** Fundamental substance or matrix of the cell (within the plasma membrane) which surrounds the nucleus, endoplasmic reticulum, mitochondria and other organelles.

**Cytotoxic:** Causing damage to cell structure or function.

**Cytochrome:** Haemoprotein whose characteristic mode of action involves transfer of reducing equivalents associated with a reversible change in oxidation state of the haem prosthetic group: strictly, the cytochrome P450 family are not cytochromes but haem-thiolate proteins.

**DDT:** A persistent organic pollutant and its degradation generally occur slowly. Breakdown products in the soil environment are DDE (1, 1-dichloro-2, 2-bis (p-dichlorodiphenyl) ethylene) and DDD (1, 1-dichloro-2, 2-bis (p- chlorophenyl) ethane), which are also highly persistent and have similar chemical and physical properties.

**Dehydrogenase:** Enzyme which catalyses oxidation of compounds by removing hydrogen.

**Dermal:** Pertaining to the skin.

**Dermatitis:** Inflammation of the skin: contact dermatitis is due to local exposure and may be caused by irritation, allergy or infection

**Detoxification:** Process, or processes, of chemical modification which make a toxic molecule less toxic.

**Dissipation:** Reduction in the amount of a pesticide or other compound which has been applied to plants, soil *etc.* (used when it is not clear whether this is by mineralization degradation, binding, or leaching).

**Dosage:** Dose expressed as a function of the organism being dosed and time, for example mg/ (kg body weight)/day.

**Dose:** The total amount of the compound given to or taken by an organism.

**Dose-response relationship:** Association between dose and the incidence of a defined biological effect in an exposed population

**Drug:** Any substance which when absorbed into a living organism may modify one or more of its functions. The term is generally accepted for a substance taken for a therapeutic purpose, but is also commonly used for abused substances.

**Economic injury level:** The pest population density sufficient to cause economic losses that is greater than the economic cost of the control action to reduce pest densities.

**Ecotoxicology:** Study of the toxic effects of chemical and physical agents on all living organisms, especially on populations and communities within defined ecosystems, it includes transfer pathways of these agents and their interactions with the environment

**Effective concentration (EC):** Concentration of a substance that causes a defined magnitude of response in a given system: EC50 is the median concentration that causes 50 % of maximal response

**Effective dose (ED):** Dose of a substance that causes a defined magnitude of response in a given system: ED50 is the median dose that causes 50 % of maximal response

**Effluent:** Fluid, solid or gas discharged from a given source into the external environment.

**Eggshell thinning:** The thinning of egg shells with a consequence of egg breakage or reproductive failure which is caused due to the presence of residue of DDT in the environment. The inhibition of carbonic anhydrase by DDT and its metabolites appears to play a major role.

**Emulsifier or emulsifying agent:** The surface active material which reduces the separation of droplets of one liquid in another.

**Elimination:** Expulsion of a substance or other material from an organism (or a defined part thereof), usually by a process of extrusion or exclusion, sometimes after metabolic transformation

**Endoplasmic reticulum:** Intracellular complex of membranes in which proteins and lipids, as well as molecules for export, are synthesized and in which the biotransformation reactions of the mono-oxygenase enzyme systems occur: may be isolated as microsomes following cell fractionation procedures.

**Environment:** Aggregate, at a given moment, of all external conditions and influences to which a system under study is subjected.

**Esterase:** A hydrolase enzyme that splits esters into an acid and an alcohol in a chemical reaction with water called hydrolysis.

**Excretion:** Discharge or elimination of an absorbed or endogenous substance, or of a waste product, and/or their metabolites, through some tissue of the body and its appearance in urine, faeces, or other products normally leaving the body.

**Exposure:** Contact with a substance by swallowing, breathing, or touching the skin or eyes.

**Fertility:** Ability to conceive and to produce offspring: for litter- bearing species the number of offspring per litter is used as a measure of fertility. Reduced fertility is sometimes referred to as subfertility.

**Fetus (often incorrectly foetus):** Young mammal within the uterus of the mother from the visible completion of characteristic organogenesis until birth: in humans, this period is usually defined as from the third month after fertilisation until birth (prior to this, the young mammal is referred to as an embryo).

**Food chain:** The feeding relationships between species or the transfer of material and energy from one species to another within an ecosystem.

**Fumigant:** Substance that is vaporized in order to kill or repel pests.

**Gene:** Structurally a basic unit of hereditary material, an ordered sequence of nucleotide bases that encodes one polypeptide chain (following transcription to mRNA).

**Glutathione S-transferase (GST):** The family of enzymes comprises a long list of cytosolic, mitochondrial, and microsomal proteins which are capable of multiple reactions with a multitude of substrates, both endogenous and xenobiotic.

**Gross Domestic Product (GDP):** The market value of all final goods and services produced within a country in a given period of time.

**Half-life (half-time) ($t_{1/2}$):** Time in which the concentration of a substance will be reduced by half, assuming a first order elimination process or radioactive decay.

**Harmful substance:** Substance that, following contact with an organism can cause ill health or adverse effects either at the time of exposure or later in the life of the present and future generations.

**Herbicide:** Substance intended to kill plants.

**Hormone:** Substance formed in one organ or part of the body and carried in the blood to another organ or part where it selectively alters functional activity.

**Hydrophilic:** Describing the character of a molecule or atomic group which has an affinity for water.

**Hydrophobic:** Describing the character of a molecule or atomic group which is insoluble in water, or resistant to wetting or hydration.

**Hyper-reactivity:** Term used to describe the responses of (effects on) an individual to (of) an agent when they are qualitatively those expected, but quantitatively increased.

**Hypersensitivity:** State in which an individual reacts with allergic effects following exposure to a certain substance (allergen) after having been exposed previously to the same substance.

**Immune complex:** Product of an antigen-antibody reaction that may also contain components of the complement system.

**Insect Resistance:** The genetic phenomenon to develop tolerance to doses of an insecticide which would prove lethal to the majority of individuals in a normal population of the same species.

**Integrated Pest management (IPM):** A pest management system that utilizes suitable techniques and methods against the pests in as compatible manner with the environment as possible and thus, maintaining the pest population levels below those causing economic injury.

**Immune response:** Selective reaction of the body to substances that are foreign to it or that the immune system identifies as foreign, shown by the production of antibodies and antibody-bearing cells or by a cell-mediated hypersensitivity reaction

**Inhalation:** The act of breathing.

**Inherently biodegradable:** Class of compounds for which there is unequivocal evidence of biodegradation (primary or ultimate) in any test of biodegradability.

**Insecticide:** Substance intended to kill insects.

**Intake:** Amount of a substance that is taken into the body, regardless of whether or not it is absorbed: the total daily intake is the sum of the daily intake by an individual from food, drinking-water, and inhaled air.

**Leaching:** A phenomenon of the movement of pesticides (dissolved in water molecules of the soil) through the soil rather than over the surface.

**Larvicide:** Substance intended to kill larvae

**Latent period/ lag period:** Delay between exposure to a disease-causing agent and the appearance of manifestations of the disease: also defined as the period from disease initiation to disease detection.

**Lethal dose:** Amount of a substance or physical agent (radiation) that causes death when taken into the body by a single absorption (denoted by LD).

**Lethal:** Deadly, fatal, causing death.

**Lethal concentration:** Concentration of a potentially toxic substance in an environmental medium that causes death following a certain period of exposure (denoted by LC).

**Lipophilic:** Having an affinity for fat and high lipid solubility: a physicochemical property which describes a partitioning equilibrium of solute molecules between water and an immiscible organic solvent, favouring the latter, and which correlates with bioaccumulation.

**Long-term exposure:** Continuous or repeated exposure to a substance over a long period of time, usually of several years in man, and of the greater part of the total life-span in animals or plants.

**Malignant:** Tending to become progressively worse and to result in death if not treated

**Metabolic activation:** Biotransformation of a substance of relatively low toxicity to a more toxic derivative

**Metabolism:** The conversion or breakdown of a substance from one form to another by a living organism.

**Metabolite:** Any product of metabolism.

**Microcosm:** Artificial test system that simulates major characteristics of the natural environment for the purposes of ecotoxicological assessment: such a system would commonly have a terrestrial phase, with substrate, plants and herbivores, and an aquatic phase, with vertebrates, invertebrates and plankton.

**Microsome:** Artefactual spherical particle, not present in the living cell, derived from pieces of the endoplasmic reticulum present in homogenates of tissues or cells: microsomes sediment from such homogenates when centrifuged at 100 000 g and higher: the microsomal fraction obtained in this way is often used as a source of mono-oxygenase enzymes.

**Microbial breakdown:** The breakdown of insecticide by microorganisms such as fungi and bacteria.

**Minor pest:** A pest species which is not causing too much harm to the crop yield or in other words it not causing much economic injury.

**Mono-oxygenase:** Enzyme that catalyses reactions between an organic compound and molecular oxygen in which one atom of the oxygen molecule are incorporated into the organic compound and one atom is reduced to water.

**Mutagen:** Any substance that can induce heritable changes (mutations) of the genotype in a cell as a consequence of alterations or loss of genes or chromosomes (or parts thereof).

**Mutagenicity:** Ability of a physical, chemical, or biological agent to induce heritable changes (mutations) in the genotype in a cell as a consequence of alterations or loss of genes or chromosomes (or parts thereof)

**Mutation:** Any relatively stable heritable change in genetic material that may be a chemical transformation of an individual gene (gene or point mutation), altering its function, or a rearrangement, gain or loss of part of a chromosome, that may be microscopically visible (chromosomal mutation), mutation can be either germinal and inherited by subsequent generations, or somatic and passed through cell lineage by cell division. **Neuron (e):** Nerve cell, the morphological and functional unit of the central and peripheral nervous systems

**Neurotoxicity:** Able to produce chemically an adverse effect on the nervous system: such effects may be subdivided into two types

**Non target organism:** Organism affected by a pesticide although not the intended object of its use

**No observed adverse effect level (NOAEL):** Greatest concentration or amount of a substance, found by experiment or observation, which causes no detectable adverse alteration of morphology, functional capacity, growth, development, or life span of the target organism under defined conditions of exposure

**No observed effect level (NOEL):** Greatest concentration or amount of a substance, found by experiment or observation, that causes no alterations of morphology, functional capacity, growth, development, or life span of target organisms distinguishable from those observed in normal (control) organisms of the same species and strain under the same defined conditions of exposure

**Paralysis:** Loss or impairment of motor function.

**Peroxisome:** Organelle, similar to a lysosome, characterized by its content of catalase, peroxidase and other oxidative enzymes.

**Persistence:** Attribute of a substance that describes the length of time that the substance remains in a particular environment before it is physically removed or chemically or biologically transformed.

**Pest:** Organism that may harm public health, which attacks food and other materials essential to mankind or otherwise, affects human beings adversely.

**Pesticide:** Strictly a substance intended to kill pests: in common usage, any substance used for controlling, preventing, or destroying animal, microbiological or plant pests.

**Pesticide residue:** Pesticide residue is any substance or mixture of substances in food for man or animals resulting from the use of a pesticide and includes any specified derivatives, such as degradation and conversion products, metabolites, reaction products and impurities considered to be of toxicological significance.

**Pharmaceuticals:** Drugs, medical products, medicines, or medicaments.

**Pheromone:** Substance used in olfactory communication between organisms of the same species eliciting a change in sexual or social behaviour.

**Photodegradation**: The breakdown of pesticides by sunlight. The rate of breakdown is dependent on the intensity of light, length of exposure, and the properties of the insecticide.

**Phytotoxic:** A material which causes damage to plants.

**Plant injury:** The damage caused to plant physiology by pest activity.

**Poison:** Substance that, taken into or formed within the organism, impairs the health of the organism and may kill it.

**Potency:** Expression of chemical or medicinal activity of a substance as compared to a given or implied standard or reference.

**ppb:** Parts per billion.

**ppm:** Parts per million.

**Probit:** Probability unit obtained by adding 5 to the normal deviates of a standardized normal distribution of results from a dose response study: addition of 5 removes the complication of handling negative values.

**Random sample:** Subset of a population that is arrived at by selecting units such that each possible unit has a fixed and determinate probability of selection

**Receptor:** High affinity binding site for a particular toxicant

**Repellent:** Substance used mainly to repel blood sucking insects in order to protect man and animals: also used to repel mammals, birds, rodents, mites, plant pests, *etc.*

**Resistance (in toxicology):** Ability to withstand the effect of various factors including potentially toxic substances

**Rodenticide:** Substance intended to kill rodents

**Route of exposure:** Means by which a toxic agent gains access to an organism by administration through the gastrointestinal tract (ingestion), lungs (inhalation), skin (topical), or by other routes such as intravenous, subcutaneous, intramuscular or intraperitoneal routes

**Rotary nozzles:** Normally used to apply pesticides at ultra-low volume (ULV) rates of application or achieve Controlled Droplet Application (CDA).

**Runoff:** The movement of pesticides dissolved in water or attached to soil particles over a sloping surface.

**Selection Pressure:** The pressure exerted during the natural processes which are ultimately responsible for the origin of new species and the adaptation of organisms to their environments.

**Small Power Sprayers:** Motor-driven power sprayers provide uniform pressure.

**Solutions:** The molecules of the toxicants or solute are uniformly mixed with the molecules of the solvent.

**Spreader:** Substance which increases the firmness of attachment of materials to the surface.

**Suspension:** A solid particles dispersed in a liquid but not dissolved.

**Synergy or synergism** (Greek word, synergos meaning working together): Refers to the phenomenon in which two or more or agents acting together creates an effect greater than the sum of the effects each is able to create independently.

**Toxicant:** The substance in formulation that is intended to kill the pests.

**Trombone Sprayer:** A medium-sized, hand-held piece of equipment. The intake tube of the sprayer is inserted into the mixture in the bucket and pump pressure is created by operating the sprayer in a trombone-like motion. The pesticide is pulled up the hose and gose out from the end of the sprayer.

**Secondary metabolite:** Product of biochemical processes other than the normal metabolic pathways, mostly produced in micro-organisms or plants after the phase of active growth and under conditions of nutrient deficiency.

**Side effect:** Action of a drug other than that desired for beneficial pharmacological effect.

**Symptom:** Any subjective evidence of a disease or an effect induced by a substance as perceived by the affected subject

**Synapse:** Functional junction between two neurones, where a nerve impulse is transmitted from one neurone to another

**Syndrome:** Set of signs and symptoms occurring together and often characterizing a particular disease-like state.

**Synergism:** Pharmacological or toxicological interaction in which the combined biological effect of two or more substances is greater than expected on the basis of the simple summation of the toxicity of each of the individual substances.

**Synergistic effect:** a biologic response to multiple substances where one substance worsens the effect of another substance. The combined effect of the substances acting together is greater than the sum of the effects of the substances acting by themselves.

**Systemic effect:** Consequence that is of either a generalized nature or that occurs at a site distant from the point of entry of a substance: a systemic effect requires absorption and distribution of the substance in the body.

**Target organ(s):** Organ(s) in which the toxic injury manifests itself in terms of dysfunction or overt disease

**Teratogen:** Agent that, when administered prenatally (to the mother), induces permanent structural malformations or defects in the offspring

**Teratogenicity:** Potential to cause or the production of structural malformations or defects in offspring

**Threshold:** Dose or exposure concentration below which an effect is not expected

**Topical:** Pertaining to a particular area, as in a topical effect that involves only the area to which the causative substance has been applied.

**Toxic substance:** Material causing injury to living organisms as a result of physicochemical interactions

**Toxic:** Able to cause injury to living organisms as a result of physicochemical interaction.

**Toxicology:** Scientific discipline involving the study of the actual or potential danger presented by the harmful effects of substances (poisons) on living organisms and ecosystems, of the relationship of such harmful effects to exposure, and of the mechanisms of action, diagnosis, prevention and treatment of intoxications

**Toxin:** Poisonous substance produced by a biological organism such as a microbe, animal or plant

**Transgenic:** Adjective used to describe animals carrying a gene introduced by micro-injecting DNA into the nucleus of the fertilized egg

**Uptake:** Entry of a substance into the body, into an organ, into a tissue, into a cell, or into the body fluids by passage through a membrane or by other means.

**Volatilization:** The process of conversation of a solid or liquid into a gas and thus, a dissolved sample is vaporized.

**Xenobiotic:** Any substance interacting with an organism that is not a natural component of that organism.

## REFERENCES

[1]    Dileep K. Singh (2007) Toxicology of insecticides. http://nsdl.niscair.res.in/ , CSIR, New Delhi, India.

[2]    Dileep K. Singh (2007) Insecticidal method of pesticide application. http://nsdl.niscair.res.in/ , CSIR, New Delhi, India.

*Pesticide Chemistry and Toxicology,* 2012, 135-137

# Index

**Dileep K. Singh**
**All rights reserved - © 2012 Bentham Science Publishers**

## T

## V

# Unit Conversion Table

| Standard Prefixes | | |
|---|---|---|
| **Prefix used in code** | **Prefix for written unit** | **Multiplier** |
| **da-** | deka- | 10 |
| **h-** | hector- | 100 |
| **k-** | kilo- | 1000 |
| **M-** | mega- | 1e6 |
| **G-** | gaga- | 1e9 |
| **T-** | tear- | 1e12 |
| **P-** | peta- | 1e15 |
| **E-** | exa- | 1e18 |
| **Z-** | zeta- | 1e21 |
| **Y-** | yotta- | 1e24 |
| **d-** | deci- | 1e-1 |
| **c-** | centi- | 1e-2 |
| **m-** | milli- | 1e-3 |
| **mu-** | micro- | 1e-6 |
| **n-** | nano- | 1e-9 |
| **p-** | pico- | 1e-12 |
| **f-** | femto- | 1e-15 |
| **a-** | atto- | 1e-18 |
| **z-** | zepto- | 1e-21 |
| **y-** | yocto- | 1e-24 |

| Standard Units | | | |
|---|---|---|---|
| **Unit** | **Symbol** | **Definition** | **Comments** |
| Time | | | |
| **second** | sec | 1 s | |
| **minute** | min | 60 s | |
| **Hour** | hr | 60 min | |
| **Hour** | hour | 1 hr | alternate symbol |
| **Hour** | h | 1 hr | alternate symbol |
| **Day** | day | 24 hr | |
| **Shake** | shake | 10 ns | |
| **Hertz** | Hz | $1 s^{-1}$ | |
| Length or Distance | | | |
| **international foot** | ft | 0.3048 m | |
| **Inch** | in | 1.0/12.0 ft | |

Dileep K. Singh
All rights reserved - © 2012 Bentham Science Publishers

| international mile | mile | 5280.0 ft | |
|---|---|---|---|
| international mile | mi | 1 mile | alternate symbol |
| milli-inch | mil | 0.001 in | |
| Parsec | pc | 3.085678e16 m | |
| League | league | 3 mile | |
| Astronomical Unit | ua | 1.49598e11 m | |
| Astronomical Unit | AU | 1.49598e11 m | alternate symbol |
| Yard | yd | 3 ft | |
| Angstrom | Ang | 1e-10 m | |
| Angstrom | \\AA | 1 Ang | alternate symbol |
| furlong | furlong | 220 yd | |
| fathom | fathom | 6 ft | |
| Rod | rd | 16.5 ft | |
| U.S. survey foot | sft | (1200./3937.) m | |
| U.S. survey mile | smi | 5280 sft | also called statue mile |
| Point | pt | 1./72. in | Typeface Point |
| Pica | pica | 1./6. in | Typeface Pica |
| **Temperature** | | | |
| Celsius | C | 1 K -273.15 | |
| Rankine | R | 5.0/9.0 K | |
| Fahrenheit | F | 1 R -459.67 | |
| **Mass** | | | |
| Gram | g | 0.001 kg | This is case sensitive. |
| Gram | gm | g | (alternate symbol) |
| pound mass | lbm | 0.45359237 kg | (avoirdupois) |
| Troy pound | lbt | 0.3732417 kg | (apothecary) |
| carat (metric) | carat | 0.2 g | |
| Slug | slug | 1 lb sec^2/ft | |
| Snail | snail | 1 lb sec^2/in | |
| Short Ton | ton | 2000 lbm | |
| Long Ton | ton_l | 2240 lbm | |
| Ounce | oz | 28.34952 g | (avoirdupois) |
| Grain | gr | 64.79891 mg | |
| Pennyweight | dwt | 1.55174 g | |
| **Force or Weight** | | | |
| Newton | N | 1 kg m/s^2 | |
| Dyne | dyn | 1e-5 N | |
| pound force | lb | lbm G | |
| pound force | lbf | lbm G | |

| poundal | poundal | 1 lbm ft/sec^2 | |
|---|---|---|---|
| kilopound | kip | 1000 lbf | |
| kilogram force | kgf | kg G | |
| **Energy** | | | |
| Joule | J | 1 N m | |
| British Therm. Unit | BTU | 1055.056 J | (International Table) |
| British Therm. Unit | Btu | 1 BTU | alternate symbol |
| British Therm. Unit | BTU_th | 1054.350 J | (Thermochemical) |
| calorie | cal | 4.1868 J | (International Table) |
| calorie | cal_th | 4.184 J | (Thermochemical) |
| Calorie | Cal | 4.1868 kJ | (nutritionists) |
| electron volt | eV | 1.602177e-19 J | |
| erg | erg | 1e-7 J | |
| Ton of TNT | TNT | 4.184e9 J | |
| **Power** | | | |
| Watt | W | 1 J/s | |
| Horse Power | hp | 550 ft lb/s | |
| **Pressure** | | | |
| bar | bar | 1e5 N/m^2 | |
| Pascal | Pa | 1 N/m^2 | |
| Pounds per sq. inch | psi | 1 lb/in^2 | |
| Pounds per sq. ft. | psf | 1 lb/ft^2 | |
| kilo psi | ksi | 1000.0 psi | |
| atmospheres | atm | 1.01325e5 N/m^2 | |
| inches of Mercury | inHg | 3.387 kPa | |
| millimeters Mercury | mmHg | 0.1333 kPa | |
| Torr | torr | 1.333224 Pa | |
| **Volume or Area** | | | |
| Liter | L | 1/1000.0 m^3 | |
| gallon | gal | 3.785412 L | |
| Pint (U.S. liquid) | pint | 1/8. gal | |
| Quart (U.S. liquid) | qt | 2 pint | |
| Pint (U.S. dry) | dpint | 0.5506105 L | |
| Quart (U.S. dry) | dqt | 2 dpint | |

| Acre | acre | 1/640.0 smi^2 | |
|------|------|---------------|---|
| Hectare | ha | 10000 m^2 | |
| Barrel (petroleum) | barrel | 158.9873 L | |
| Fluid Ounce | oz_fl | 29.57353 mL | |
| Gill (U.S.) | gi | 0.1182941 L | |
| Peck (U.S.) | pk | 8.809768 L | |
| Tablespoon | tbl | 1/32. pint | |
| Teaspoon | tsp | 1/3. tbl | |
| Cup | cup | 16. tbl | |
| **Electromagnetism** | | | |
| Coulomb | Co | 1 A s | Electric Charge |
| Volt | V | 1 W/A | Electric Potential |
| Ohm | ohm | 1 V/A | Electric Resistance |
| Ohm | \\Omega | 1 V/A | alternate symbol |
| Faraday | faraday | 96485.31 Co | Electric Charge |
| Farad | farad | Co/V | Capacitance |
| Stokes | stokes | 1e-4 m^2/s | |
| Oersted | Oe | 79.57747 A/m | |
| Webber | Wb | V s | Magnetic flux |
| Tesla | Tesla | Wb/m^2 | Magnetic flux density |
| Henry | H | Wb/A | Inductance |
| Siemens | S | A/V | Electrical Conductance |
| **Light and Radiation** | | | |
| Lux | lux | cd/m^2 | Iluminance |
| Lux | lx | cd/m^2 | |
| Lumen | lm | cd | Luminous Flux |
| Stilb | sb | 10000 cd/m^2 | |
| Phot | ph | 10000 lx | |
| Becquerel | Bq | s^-1 | Activity |
| Gray | Gy | J/kg | Absorbed Dose, kerma |
| Sievert | Sv | J/kg | Dose equivalent |
| **Other Quantities** | | | |
| pound mole | lbmole | 1 mol lbm/g | Quantity |
| Poise | poise | 1 g /sec cm | Viscosity |
| Gravity's accel. | G | 9.80665 m/sec^2 | Gravity on Earth |
| Degree | deg | Pi/180 | Can be used to convert from degrees to radians for trig functions. |
| Percent | % | 0.01 | |

| Knot | knot | 1852 m/hr | Velocity |
|---|---|---|---|
| **Miles per Hour** | mph | 1 mi/hr | Velocity |
| **Gallon/minute** | gpm | 1. gal/min | flow rate |
| **Revolution/minute** | rpm | 360 deg/min | |

Source : http://www.csgnetwork.com/converttable.html

www.ingramcontent.com/pod-product-compliance
Lightning Source LLC
Chambersburg PA
CBHW041713210326
41598CB00007B/634

* 9 7 8 1 6 0 8 0 5 5 3 1 9 *